UNDERSTANDING GENETICS

STRATEGIES FOR TEACHERS AND LEARNERS IN UNIVERSITIES AND HIGH SCHOOLS

Anthony J. F. Griffiths
Jolie Mayer-Smith
University of British Columbia

W. H. Freeman and Company
New York

ISBN 0-7167-5216-6

Printed in the United States of America

Second printing, 2001

Contents

Preface

This book is the product of eight years of collaboration between the authors, who are based in the faculties of science and education, and who share a common interest in improving teaching and learning in biology, specifically in genetics. The ideas presented here were generated through lengthy periods of observation, discussion, and analysis of problems instructors face when trying to assist their students in achieving a deep understanding of the principles of genetics. The strategies and activities that follow come from working directly with students enrolled in genetics courses at our university and build upon the collective wisdom of former teaching assistants, staff, and faculty who participated in Collaborative Study Group Meetings in 1993-1994, 1994-1995, and 1997-1998.

Some words of advice and caution. First, we have learned through the hard hand of experience that many instructional strategies and teaching procedures are by nature and design contextually bound. This means that application of the ideas presented here should not be followed in a recipe-like fashion, but rather will be most useful if they are modified for one's particular instructional situation and group of students. Instructors will need to gauge their individual groups of students to determine when and where to use each particular strategy as they proceed through a genetics course. Second, it is particularly important to realize that adopting and using a new teaching strategy requires the acquisition of new skills by both instructor and students. Instructors should allow some time to introduce any new teaching and learning procedure and not become seriously discouraged if the new strategy is not immediately understood or embraced by students. Finally, because students usually arrive in classes with firm ideas about what constitutes legitimate teaching and learning practices, they may disregard a valuable new strategy that requires them to change their views or behavior. Therefore, it is essential for instructors to explain why a new or different teaching procedure is being used and how it will help students with their learning. That is, they should listen to and address their students' concerns. Whereas this may not entirely alleviate student reservations, it may help them begin to think about how one's learning approach can affect one's understanding of a subject.

The ideas presented in this book are based upon five fundamental beliefs about teaching and learning genetics.

1. Students find genetics to be one of the most difficult life science subjects to master. Conversely, genetics is not an easy subject to teach.

2. All students are capable of mastering genetics, but some do not have the necessary skills at the beginning of a genetics course, and others never acquire these skills.

3. Traditional teaching and learning methods do not seem to be effective in helping students to learn genetics, despite their effectiveness in other science subjects.

4. The contributory causes of learning difficulties are
   (i) Students' beliefs about the role of the instructor and the student in a learning situation.
   (ii) Student passivity, resulting from their previous (apparently successful) educational experiences.
   (iii) Instructors' beliefs about what teaching practice helps students learn, namely that clear presentation of material is an adequate instructional practice.

5. To alleviate these difficulties it is necessary to provide more opportunities for structured, active, and constructive processing of new information accompanied by some discussion of instructional beliefs and practices.

The rest of this book is devoted to discussing student learning problems and describing methods for helping students improve their ability in the areas of information processing and self-monitoring of understanding. These skills are key to achieving an improved and deeper understanding of genetics principles, which are in turn the key to scholarship in biology and to many of today's pressing social issues.

We are indebted to Dr. Barbara Moon of the University College of the Fraser Valley, who read and improved most of these chapters. We also acknowledge the participation of the teaching assistants and students of Genetics I and II at the University of British Columbia, who acted as guinea pigs for most of our ideas, helped us refine them, and also gave us many new ideas.

Drafts of most of the chapters were published by the authors as regular teaching and learning columns in the *Bulletin of the Genetics Society of Canada*, and we are grateful for the agreement of the Editor to publish them in the present form.

1

CONSTRUCTIVISM FOREVER!

The conceptual underpinning for most of the ideas on teaching genetics in this book stems from a theory of learning called constructivism, which has gained wide acceptance and become a paradigm for educational theorists and practitioners. Although many of the tenets of constructivism will seem logical and "obvious" at first glance, their application to teaching practice challenges what many of us have traditionally regarded as fine teaching. We believe that to adhere to constructivist principles will require a large change in the teaching style of most practitioners. Furthermore, we believe this change can radically improve the learning that goes on in our classrooms. We'll say more about constructivism later, but first a digression. Given that we are writing about moving away from "traditional" teaching practice, how might we characterize this traditional approach? Here are a few of its characteristics.

Traditionally:

The role of a teacher is

- to tell.
- to be in control of the pace and content of classes.
- to be the purveyor of the truth, knowledge, facts, wisdom.

The role of a student is

- to be the passive recipient of knowledge.
- to be dependent upon the teacher.
- to be non-reflective.

As a result the student in a traditional setting may demonstrate any of the following learning tendencies:

- accepting information uncritically
- inability to transfer learning to new situations
- poor skills at problem solving

This approach has been called "mug and jug education." The professor represents the jug, brimming full with knowledge and expertise, and it is this knowledge that is poured out into the passively recipient vessels, the mugs, represented by the students.

A different perspective on learning, with highly significant implications for teaching genetics, is based on an extensive body of research that has been developed over the past two decades (since the early 1980s). This perspective is called constructivism.

In simplified terms constructivist theory is based on the idea that learning involves the construction of knowledge by the individual student as he or she engages thoughtfully with information. Thus knowledge is not a commodity that can simply be transferred from teacher to student but rather is a product of the student's processing and making connections between new information that is received and knowledge that has been previously "constructed." Prior knowledge and experiences are very important because they serve to mediate how the student views, accepts, processes, and constructs new understandings. The role of teaching in a constructivist classroom is to provide experiences that will assist the learner in making "useful" (in our case, scientifically correct) constructions.

We reiterate that constructivism is *not* a method of teaching but rather a view of how learning occurs. It has implications for the roles of both the teacher and the student. This is particularly applicable in difficult subject areas such as genetics, in which students have to do a great deal of intellectual processing to internalize the new material. Teaching that is consistent with a constructivist perspective of learning is often referred to as "teaching for understanding and conceptual change," based on the view that students

2

enter our classrooms with an initial conceptual understanding of the world that will change as a result of the learning that takes place.

When teaching for conceptual change under the constructivist paradigm:

The role of a teacher is
- to facilitate intellectually active learning.
- to share control of learning.
- to promote student thinking about concepts and learning.

The role of a student is
- to be an active constructor of knowledge.
- to be independent and purposeful.
- to be metacognitive—i.e., to think about their thinking.

As a result, the student may demonstrate the following learning tendencies:
- critical thinking
- skill at transferring learning to new situations
- skill at problem solving

The strategies we introduce in the chapters that follow are based upon this particular perspective on learning.

When implementing any type of change at an institutional and at a personal level, individual epiphanies take place, during which the need for change is established beyond doubt. Anecdotes from our own individual intellectual "journeys" demonstrate how each of us came to question and challenge our classroom practice.

Pondering the sticky business of learning science

When I began teaching in 1981, I viewed it as a "sharing" of scientific knowledge. I seemed to be highly successful and regularly received praise and positive feedback from my students. But it gradually dawned on me that in this "shared enterprise" I was doing more work and learning more than my students. The students were interested and attentive, but while I was extremely active, they were frequently very passive. A number of them couldn't seem to learn what I was teaching. I tried to shift the focus of lessons onto the students. My lectures contained more discussion questions and examples. I became adventuresome and tried to implement creative approaches and more "hands-on" techniques, including problem solving during labs, student-directed projects, group work, and model building. Still I noticed that quite a few seemingly dedicated students had a lot of trouble learning specific topics in the biology curriculum.

One particularly difficult topic area for biology students was related to cell function. Having a background in cell biology, this particularly intrigued me. Students would seem to learn the concepts, but they had a lot of trouble dealing with questions that required them to apply what they had learned in a new situation. In one instance I worked through a particularly complex biology unit on cell processes and transport that included issues of concentration gradients, osmotic pressure, and movement of molecules across membranes. To assist with the abstract nature of the concepts we held class discussions, drawn pictures, built models, run a lab, and practiced answering application questions concerning the survival of cells and organisms in solutions of different concentrations. Satisfied that I had dealt with all possible causes of student confusion, I set an exam that balanced recall with thought.

I settled back to read students' responses to the following application question: *Why doesn't honey spoil (become contaminated) even when exposed to bacteria? Use cell transport principles in your answer.* The cell transport principles we had discussed in detail were listed in a question on the next page of the test. As I read their responses I became more and more shocked to discover that only ten percent of the students in my

4

classes had given a scientific explanation that discussed how the osmotic pressure exerted by the sugar molecules in the honey would tend to draw the water out of the bacteria cells. The remainder had given one of two explanations.

1. The bacteria got stuck and trapped by the sticky honey and starved to death. (A "sticky view" of honey as a germicide.)
2. The cells in the honey were too close together and the bacteria couldn't squeeze in between them. (A "cellular view" of honey as a germicide.)

I was horrified! My students were telling me things I would expect to hear from elementary children. It was as if the three weeks of rigorous teaching of cell processes had never taken place. I had read about young students having prior and naive views of science, nature, and the world that influenced their learning and understanding of scientific principles, but I never expected to find this problem emerging among my "mature students." But here was crystal clear evidence of what the literature referred to as "children's science." This was the critical incident that made me dramatically change the way I looked at teaching and learning. (Personal anecdote, J. M-S. 1992)

A "convert" to constructivism

Six years ago, my position on teaching and learning was that I did the teaching and the students did the learning. I felt that it was my job to deliver the material of genetics from my informed standpoint in as clear and inspiring a way as I possibly could, and it was the student's job to wrestle with the concepts and demonstrate a working understanding of them by doing problems. The trigger that prompted me to change my thinking was that whereas I seemed to be getting better at my job, the students seemed to be getting worse at theirs. In an attempt to keep the marks in the course up, I had started making the exams easier and easier, but still the marks were not responding. I read about data that showed that this is a North American trend, with performances in many academic tests declining. My geneticist colleagues agreed with my assessment that the situation was particularly bad in this discipline.

It then occurred to me that (a) for many students, my being very clear *wasn't* enough; and (b) many of the students' difficulties were not their fault, but rather a reflection of their prior exposure to science teaching methods and other broader public trends. Exploration of the matter with colleagues in education led to my reading selected literature, which suggested that the problem was most likely student passiveness in learning. Furthermore, interviews with high- and low-scoring students in my course suggested that the major difference between them was that the high scorers were able to assess their analytical difficulties with an internal check system and then respond by changing their learning methods from regurgitation mode (useful in other courses) to a metacognitive or self-questioning mode. The above ideas convinced me that a genetics teacher has an obligation to help students become metacognitive, and that this cannot be done effectively with traditional modeling methods of teaching. Methods are needed to jump-start the students into processing and active engagement in dialogues about the material. It is now my belief that this is the only way that students can deconstruct their previous genetic conceptions and traditional learning methods and reconstruct for themselves the mental schemas needed for genetic analysis. Admittedly, some can do this without our help, but many cannot. In short I have been converted to the constructivist philosophy of education. I must admit that there are few data that bear either way on the success of this approach, but I am convinced that traditional methods are inadequate for a sizeable proportion of students, so in the absence of any better philosophy, I am personally committed to giving constructivism a try. (Personal anecdote, A. J.F. G. 1992)

Perhaps these two anecdotes have prompted you to reflect on your own experiences with learning and teaching. Have you ever been surprised at the views of your students or colleagues? Have you ever wondered why a concept was so difficult to learn or teach?

So, what are some of the basic tenets of constructivism and how can it inform educational practice? Let's take a closer look at the constructivist perspective and its implications.

A constructivist perspective of learning: Some fundamental beliefs

1. Learners have prior beliefs and understandings that may influence how they learn new concepts.

2. Learners' beliefs and understandings may be deeply ingrained, strongly held, and persistent—therefore difficult to change.

3. Learning involves the linking/connecting of new concepts with prior understanding and ideas, and this will require learners to

   • add to/enhance their previous store of ideas = **construction.**

   • replace prior views and ideas with more plausible, useful ones = **reconstruction.**

4. The **construction** or **reconstruction** aspects of learning occur most readily when learners think about

   • what they do and don't understand.

   • how new information fits with what they already know.

   • what they are doing and why they are doing it.

5. Thus, **construction** and **reconstruction** require that learners engage in **self-monitoring** of

   • personal understanding.

   • their own performance as it compares with a) intent of the learning experience and
      b) instructions pertaining to the learning experience.

In summary, according to a constructivist view of learning, learning involves

- **Construction**
- **Reconstruction**
- **Self-Monitoring**
  - of personal understanding
  - of performance against
    - intent
    - instructions

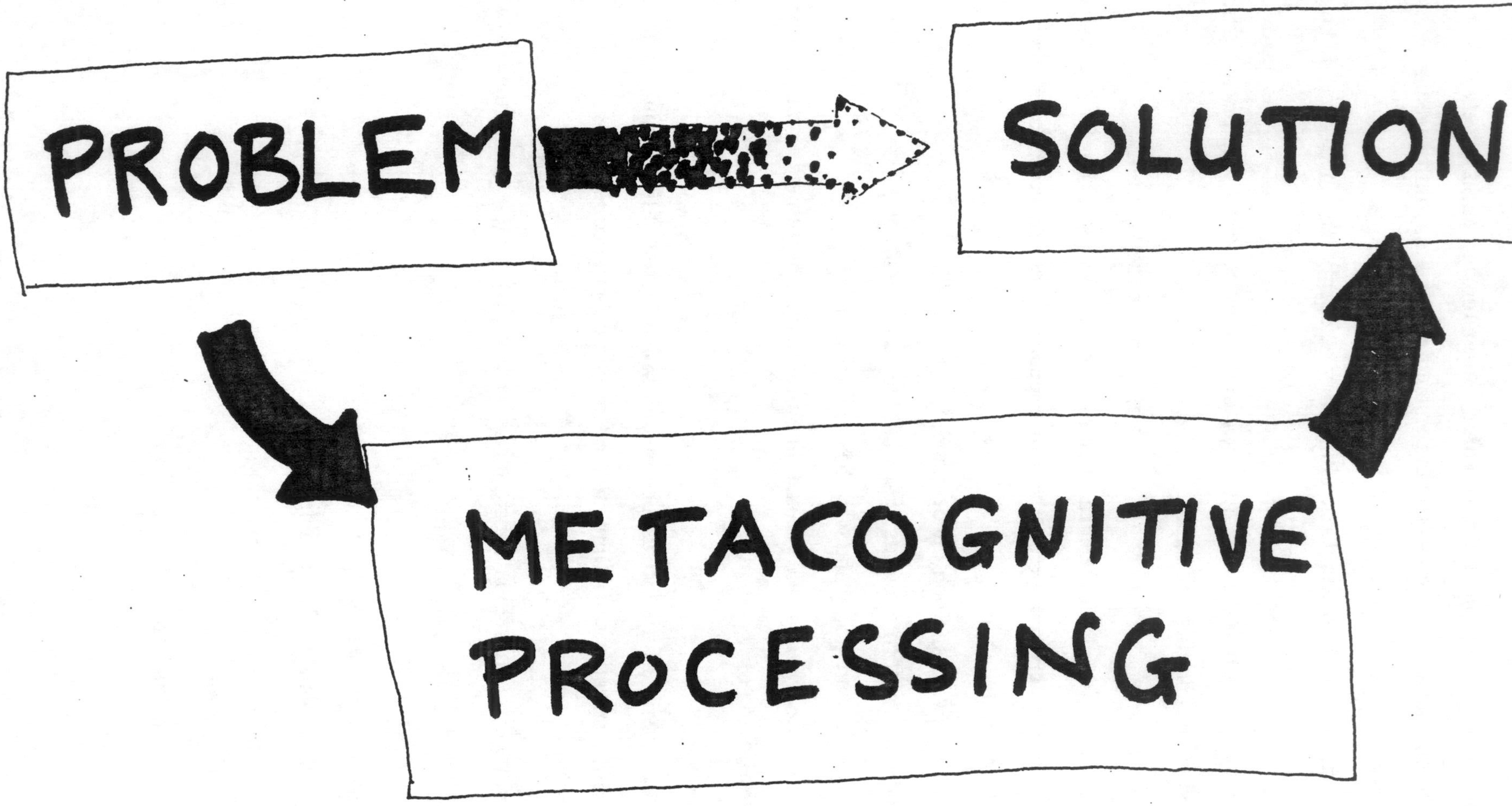
PROBLEM
SOLUTION
METACOGNITIVE PROCESSING

Obviously a class that focuses on teaching for conceptual understanding requires an environment that is rather different from that found in a traditional teaching and learning setting. These differences pertain to the types of behaviors and attitudes of both the instructor and the students in the class. Below we list some distinguishing characteristics of such a class.

A class environment conducive to teaching for conceptual understanding

Characteristics:

1. Student support and consent are involved.
2. Risks are taken by students and instructor.
3. Collaboration and negotiation about pace, concepts, and activities occurs.
4. Ownership and input are shared—students' ideas and questions are listened to and used.
5. Student talk is tentative, hypothetical, exploratory.
6. Students change
    - learning behaviors.
    - conceptions about learning.
    - attitudes about learning (roles of instructor and student).
    - trusts.
7. Student assessment is consistent with the approach to learning.

By now you may be asking yourself: What is the big problem? Why bother trying to change the beliefs and behaviors of our learners? Don't passive learners succeed just fine? Perhaps in some courses they do, but from many years of experience we are convinced that traditional instructional practices and passive learning behaviors add up to failure when it comes to learning and being able to apply genetics principles. And, although we haven't emphasized this yet, there are many additional educational spinoffs that emerge from assisting our students in moving from being passive learners to becoming thoughtful, reflective, metacognitive learners. The table below may help convince you of some of these benefits. It is from *Learning from the PEEL experience*, Edited by J. R. Baird and J.R.Northfield (1992).

Contrasting student beliefs about learning and teaching

The Passive Learner Believes	The Metacognitive Learner Believes
1. School learning is very short term; for the period up to assessment.	School learning is about permanent changes in skills and understandings.
2. The goal of school learning is satisfactory assessment. Something that isn't to be assessed is of very little value or interest. Once it is assessed it is of no further interest.	The goals of school learning include satisfactory assessment, but a stronger motivation is the satisfaction of genuine understanding. Assessment has an important formative role.
3. School learning comes in independent packages. The ideas or skills in one topic or subject will very rarely relate to or be useful in another.	The ideas or skills from one topic or subject ought to be consistent with and may well be useful in another.
4. Knowing the purpose of an activity and its links with earlier lessons is irrelevant. The goal is to follow instructions taking the minimum time.	Knowing the purpose and links are crucial to genuine learning.
5. Learning involves remembering, not independent thinking, hence assessment should require reproduction of what has been specifically covered in class.	Learning requires independent thinking and hence assessment should include new tasks requiring new applications of general principles or procedures.

6. The reasons for answers or for steps in procedures are of little value or interest.

The reasons are essential for building understanding and need to be searched for.

7. Real work involves writing (often copying). Discussion and thinking are not real work.

Copying is low-grade work. Discussion is difficult, intellectually challenging work.

8. Exploring wrong or alternative answers is of no value. Similarly, explaining things in two ways is confusing. Confusion is of no value to learning.

Exploring wrong or alternative answers is frequently very helpful as is explaining things in two ways. Periods of confusion are often both necessary and helpful to learning.

9. To admit to not following any part of a lesson or explanation is to admit to being stupid.

To be able to identify the ideas or parts of a lesson which don't make initial sense is a sign of progress.

10. Teachers are entirely responsible for each student's learning. This means that their job is to clearly present all the right answers. The student's job is to memorize these.

Each student has a major role in and responsibility for their own learning. The teacher's job is to help them construct genuine understandings.

11. The teacher's answers ought to be accepted uncritically. To challenge them is a waste of time.

New ideas should be examined critically. Disagreement or challenge often helps understanding.

12. Students' current beliefs are irrelevant.

It is important for students to retrieve and reflect on their views, to contrast them with those of other people.

11

| 13. | The teacher is entirely responsible for the classroom activities. Teachers set tasks and ask questions. Students perform those tasks and answer questions. | Students regularly collaborate with the teacher and influence the classroom activities, often by raising important questions that require answering. |

Reproduced by permission of the authors.

Many students need no assistance in becoming metacognitive, active learners—for some reason it comes naturally to them. However, it is our experience that most students cannot do this and need help. This is the job of the professor, whose role under the constructivist approach becomes that of a learning facilitator. The facilitator's role is also active but in a different way. He or she must design and provide learning activities that encourage active processing.

The purpose of these processing tasks is to encourage students to engage actively with their curricular material. From experience we have found that a large impediment to learning genetics is that students attempt to apply passive learning techniques that have worked successfully for them in many other courses. This approach does not succeed in promoting their learning of genetics because genetics is an analytical, problem-solving discipline requiring active engagement with data interpretation.

In the chapters that follow we tell more of our stories and share a wide variety of strategies we have tried with students and instructors of genetics. We have found these useful and we hope you do as well.

Reference

Baird, J.R., & Northfield, J.R. (Eds.). (1992). *Learning from the PEEL experience*. Melbourne, Australia: Monash University Printing Services.

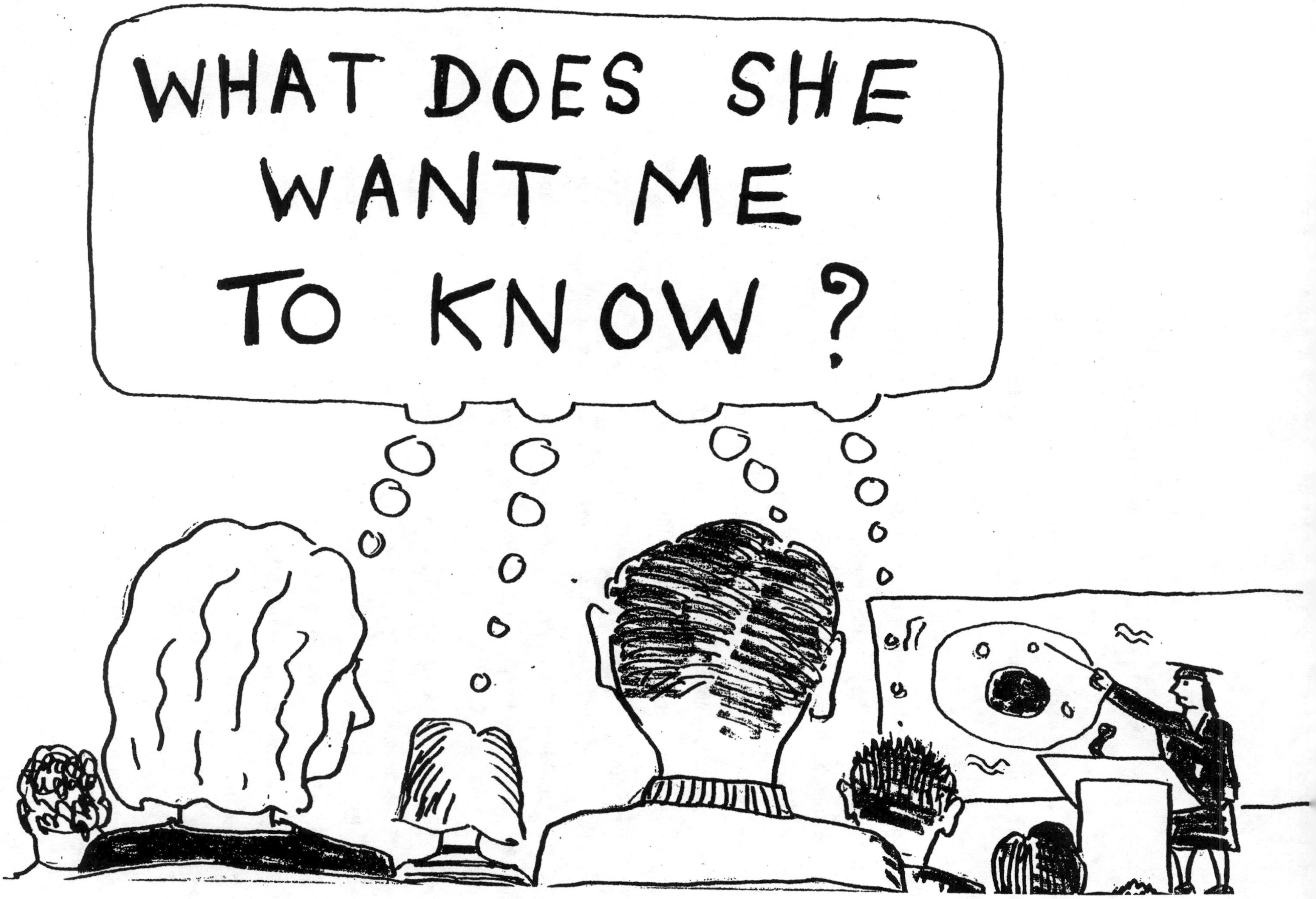
WHAT DOES SHE WANT ME TO KNOW?

2

The End of Lecturing as We Know It?

Does lecturing have a place in today's world of the Internet, CDs, and sophisticated textbooks? In this chapter we explore the role of the lecture in modern genetics teaching and learning.

A few years ago we attended a workshop on the subject of teaching large classes. At the workshop we waited with eagerness for the leader to tell us something new that might overcome this challenging aspect of mass education. However, we were disappointed. We heard suggestions like: "...make sure you know your stuff, make sure you are organized, check your equipment before the lecture, and don't indulge in distracting activities like head scratching, harrumphing, coin-jingling, etc." Although these are important issues to keep in mind when teaching any class, they do not address the substantive issues of what teaching and learning strategies are most suitable for the masses, and how we can assist our students in acquiring a deep understanding of our subject matter. Whereas we realize that there are no simple answers to these questions, our quest for solutions has provided us with some insights that we share in this chapter.

Historically, university lectures evolved at a time when there was inadequate access to books. The lecturer was the only font of wisdom and students crammed themselves into lecture halls because they really had no other choices. Yet today, with excellent libraries, textbooks, and Internet resources available to most students, they are still cramming themselves into overcrowded lecture halls. One might expect that lecturers, although perhaps not in the same league as the dodo, should be at least a dying breed. How can we explain this paradox, this perennial popularity of the lecture?

There are numerous possible reasons—many related to lecturing in general and not specifically to genetics instruction. Undoubtedly the inspirational potential of the lecture is one of its major assets. To see and listen to a professional practitioner of genetics expound his or her expertise and convey why genetics is so appealing has a magnetism that is difficult to muster on the printed page or on the computer screen. Then again the ability to unify the subject, tie together threads from various topics, and make the subject seamless and self-consistent is a talent that is worth its weight in gold, and another aspect that is much more difficult to convey through other teaching approaches or media. This *integrative* feature is particularly important in genetics because of the difficulty students have in determining which genetics mechanisms (e.g., crossing over, translocation, mutation, or nondisjunction) best explain the significance of a given set of experimental data.

Unfortunately and regrettably many university lectures seem to lack both these appealing features. In the lecture hall we find the professor bogged down in the swamp of detail, a devotee of an invisible curriculum deity who mandates that learning science is synonymous with memorizing a mountain of material rather than learning how to be a questioning and creative scientist. For their part, we see the students working frantically during lectures, meticulously copying down notes on material that is present in their textbooks with hardly a thought about what it all means. No wonder ninety-nine percent of what we teach our undergraduates is forgotten shortly after their examinations are written. Our devotion to lecture-based instruction may also explain why successful undergraduates who enter our graduate programs arrive with limited ideas of how to approach a practical scientific problem.

In courses organized around simple memorization and recall of information, the lecture format appears to be a reasonably enjoyable and painless way for "lecture-learner" students to accumulate material. Pragmatically speaking, such courses are also the easiest to teach and assess. Memorization and regurgitation are among the types of activities that the students have been reared on ever since high school. For many students accumulating a mass of stuff about science *is* science, and the scientist is someone who knows a lot of stuff. And why not?—this has been the role model presented to them.

But do students gain a deep understanding of the science concepts that they memorize in such courses? The answer to this question only emerges at exam time if students are asked to perform a simple application or solve a thought-provoking practical question. In many cases the results are stunningly depressing: students can repeat back what they have memorized but many have minimal ability to apply these ideas. This problem becomes particularly acute in genetics where most if not all the questions we ask require the application of concepts in problem-solving situations and analysis.

To compound this problem, it is clear from our interviews with students that many of them rely heavily on lectures for their information. In fact, as we all know, the question "Do we have to read the book?" is a common one. Indeed, one A-grade student in our third-year genetics course this year confessed that he has neither bought a textbook in his time at the university, nor had he felt the need to use one. But in this area of textbook reading, genetics is different, as it is in so many other areas. A common complaint is that the lectures focus on "theory" whereas the exams are concerned with data-based problem solving. The two domains of "theory" and data analysis are seen as unconnected, even though the lecture and the textbook generally contain both "theory" and analysis in close physical proximity. The lecturer's viewpoint is "I provide you with the basic concepts and analytical techniques and then you will demonstrate your understanding by doing some data analysis, i.e. doing some problems" (which are usually found at the end of the chapter in the text). The lecturer sees the connection as obvious, but in many cases it is not. In this book we will suggest a number of ways that students can address this basic problem, but first let us see what the instructor can do to help during the lecture.

First the lecturer can alert the students to the apparent theory–practice gap at the beginning of the course. The point can be made that there is a connection between the lecture material and assessment which requires application of this material, but that the link is not automatic—it is directly related to the problems being assigned. We must tell our students that lectures should be regarded as necessary but not sufficient. Point out that just sitting in the lecture and imbibing every word will certainly familiarize them with the material, but will not on its own help them apply the knowledge to a simulated real-life situation of data analysis. The students themselves must forge the connection, and this can only be achieved by their constructing a rich picture of the topic in their own terms. Only then will their understanding be in a form that they can access and use effectively for problem solving. In other words, mental work and *processing* is necessary to convert learning *about* genetics to actually *doing* genetics. Furthermore, this work is of a special kind not experienced or practiced in most other biology courses.

Another point worth making early in the course is that studies have been conducted that show that students who *process* information *during* lectures or *soon after* lectures are more successful on exams. The research compared three methods of study: (i) note-taking during lectures, and reviewing before the exam; (ii) summarizing lecture notes in one's own words after the lecture; and (iii) self-questioning during lectures. The latter group performed the best of all on examinations, and the traditional learners in the first group came last in the assessment.

Self-questioning is a method for processing and internalizing the concepts being presented during the lecture. From the students' perspective this seems the riskiest of the three methods because they are afraid they are going to miss something by not writing down everything the lecturer says or writes. Ironically in rote copying they *are* missing a great deal of processing opportunity. The self-questioning student asks and answers in his notes such questions as "What is she (the professor) trying to show here?", "Why is she using this organism? These phenotypes?", "What do these numbers really mean?", or "How does this stuff relate to the stuff we covered yesterday?" The method forces students to make links, process information, and to be metacognitive, that is to reflect on their thought and learning processes. The reflection process helps students internalize the material effectively at a crucial time, when they are actually receiving the information. This strategy leads students to adopt an active questioning approach rather than a passive, accepting role in learning. This approach is particularly well suited to genetics because of the processing link referred to above.

The summarizing method is also commendable. In this approach students take their copied lecture notes and process them on their own time, asking questions such as "What one sentence would express the main idea of this point in the lecture?" or "What is a related idea, and can I link the two ideas together in one sentence?" or "Can I draw a concept map of this lecture and related topics?"

The professors can promote and assist students in their processing by pausing during their lecture for reality checks. Here the lecturer designs and administers small processing tasks. For example, you might ask the class to make some small calculation or deduction during the lecture or ask the students to relate the topic at hand to a topic discussed two or three lectures or even weeks ago. This can be handled in a variety of ways. One method that works well to get students thinking is the dyad method. In this method, the students form dyads (pairs) with the student sitting next to them, and discuss the problem posed by the lecturer. The device works because it allows processing and also gives students practice and the confidence to speak out during the lecture. It is also an excellent device in graduate lecture courses in which students are looking at complicated primary research data. We have tried dyads in large genetics lectures of up to 200 students. Its success requires some faith and bravery on the behalf of the lecturer, careful planning, clearly stated directions and objectives (e.g., set time limits and establish a signal to gain student attention), and lots of student cooperation. An alternative strategy that we have used is to give the students individually a minute or two to ponder a small task. Then provide an answer and respond to student questions before moving on. Even this simple procedure allows for the important reality check to be made (i.e., are students following the material in a manner that will allow them to use the information received). This procedure has the added advantage because it demonstrates that you, the lecturer, attach great importance to the processing procedure.

Of course, just telling the students that a device works is not going to convince them to use it. Somehow you have to get them to give it a try. The cynic in us says that students will never try anything unless it is going to be on the exam, so why not insert a processing exercise as a small part of the exam? For example, try an expansion exercise (see Chapter Nine). This will not only further demonstrate your commitment to the process but also meet the educational goal that assessment should be congruent with instruction. Who knows, perhaps the results of that exam will show significant gains from the exercise?

Unless policies change, large classes will remain among the traditions and banes of the modern university. Those of you reading this book probably recognize that the lecture hall is neither conducive to student–lecturer interaction and feedback, nor does it provide a physical environment that easily supports contemplative tasks on difficult material. However, viewed on the positive side, perhaps lecturing to huge classes is an opportunity to make a correspondingly huge difference in teaching and learning genetics. Over the years, Tony has been regularly surprised by students coming into the office for help and starting off with a question along the lines of, "Are you Dr. Griffiths? I'm a student in your genetics course but I didn't recognize you because I sit at the back of the lecture hall." Perhaps realizing that to our beloved students our image may be exceeding small and blurred is just what is needed to move us into taking action and changing how our students view our role and their role during lectures.

We close with some ideas to give meaning to the lectures. Whether your audience is large or small, it is important that

1. Lectures should not duplicate the textbook as a means of information delivery.

2. Lectures should superimpose a layer of organization on top of the material covered by the textbook, a layer that emphasizes integration and synthesis. This should be the lecturer's unique view of the world of genetics, and should be as inspirational as possible.

3. A portion of the time spent in lecture should be set aside for students to process new data and relate it to old.

4. Students should be informed about and given examples of alternative approaches to note-taking and copying.

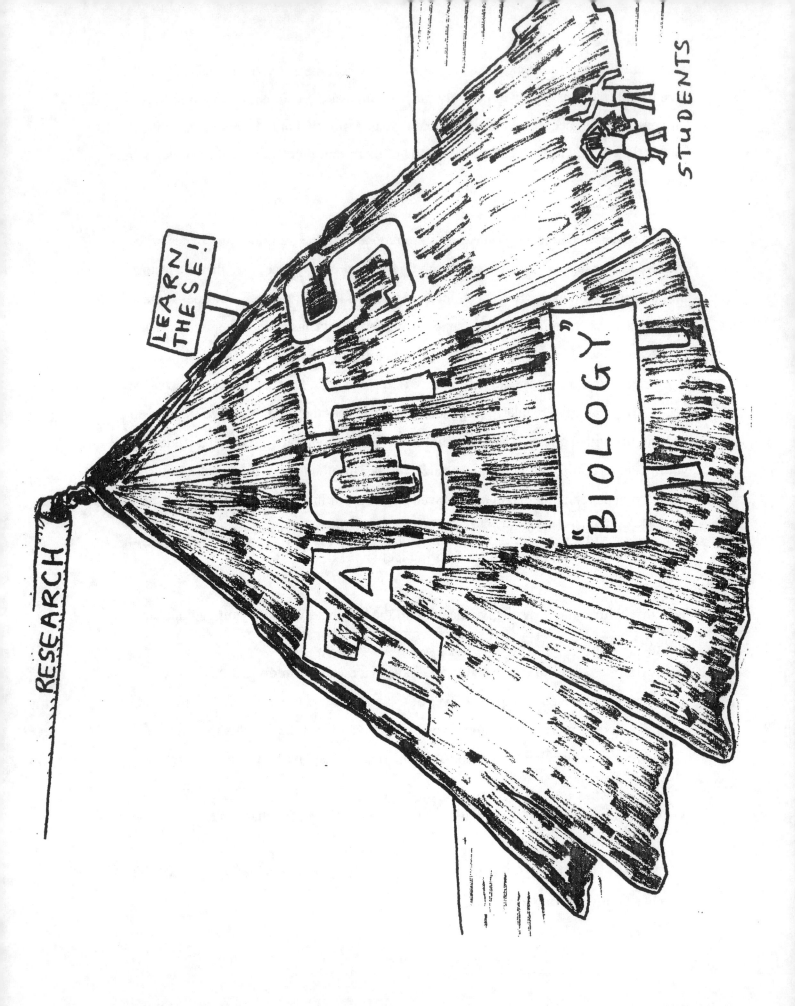

One of the most conspicuous differences that students see between high school and university or college is the presence of lecture halls. Therefore the students come to the inevitable conclusion that a lecture is a high-level university learning experience. Clearly if the university has taken a lot of trouble to build and equip these halls, the student feels they must be effective. Students expect to be lectured to; it is what they paid good tuition fees to experience. Indeed the administration would probably endorse this view. Yet lectures, as we have shown, are a medieval phenomenon that has outlived its usefulness. If lectures must be used, they must be redesigned to promote active learning, not solely for information transfer.

3

Think Aloud Problem Solving in Triads

Most undergraduate students find genetics challenging and difficult. Commonly, students who have traditionally done very well in their studies perform poorly in (or fail) their first course in genetics. A frequently heard complaint is that whereas in most courses the mark obtained is proportional to the amount of effort expended in study, in genetics there is no such relationship. There is agreement among students that genetics somehow is "different." Because this problem permeates most institutions, the villain is most likely not just the professor, but rather, a combination of the nature of the subject and how it is taught. So whereas we concede that the instructor has an important role in student achievement, what it all boils down to is problem solving. Most students are bad at it, find it frustrating, and usually end up hating genetics because of it.

However, it seems that professors the world over doggedly stick to the problem-solving method of evaluating students in genetics. Why do they do this? The reason is that the genetics problem in a small way effectively simulates the act of doing genetics. In the problem an experiment is described, results are given, and the student is challenged to interpret the data using the genetic principles learned in the course. As professionals, this is the very type of analysis we encounter every day in the lab. There may be a "best answer" to the problem, but nevertheless the analysis is generally open-ended and could go anywhere the problem solver may direct it. Often several different valid interpretations are possible. This type of exercise also comes closer to testing an understanding of science than any evaluation method based on writing short answers or essays, which, in reality few scientists ever write in their professional careers. The genetics problem provides a taste of the ultimate criterion of the worth of knowledge in science, which is whether or not the knowledge can be used. Knowledge that cannot be used is little more than glorified *Trivial Pursuit*.

One fundamental reason that students have so much trouble with problem solving is that generally they have not been educated to think as scientists. They have been trained to assimilate, and then they are evaluated on their ability to regurgitate the assimilated material. Thus, their prior experience in science classrooms affects the way students try to learn problem solving. A common situation in university and college courses that require problem solving is the tendency for students to assume a passive role and expect the instructor to show them how to solve problems. Skilled as "lecture-learners," these students believe that watching, listening, and taking notes while an expert models how they go about their profession will help them to become proficient. Even more troubling is that in small group tutorial settings, which are designed to encourage student *participation*, most students prefer to sit and *watch* their instructors give mini-lectures and model more problems. Unfortunately, experience shows that these methods do not work for problem solving. At exam time, even the slightest curveball in a question can throw off students who have tried to learn genetics problem solving by modeling.

Having observed these characteristic passive-learning tendencies among students enrolled in a third-year required genetics course at the University of British Columbia, we set ourselves the objective of finding a strategy to promote more active student participation during small group problem-solving sessions. We introduced a teaching procedure that could be used as a tool by the graduate teaching assistants who present the tutorials in this course. The procedure we introduced is a modification of an instructional strategy called TAPS (Talk Aloud Pair Problem Solving) first developed by Whimbey (1984) and adapted by Pestel (1993) for use in university chemistry classes. In Pestel's TAPS procedure, pairs of students are assigned speaking or listening and facilitating roles, and engage in solving problems while discussing their thinking aloud. We modified Pestel's approach by adding a third member to the group who acts as a scribe. This individual keeps a record of the issues discussed, the points that remain unclear, and instances where more information is needed. Including a third student offers the advantage of providing an avenue whereby the important issues discussed by the pairs of students are not lost at the end of the small group activity, and establishes a forum for subsequent discussion by the entire class.

In brief, the method is as follows. The class is given a new genetics problem that should be solvable using the principles or concepts previously introduced in the course. This might involve the analysis of quantitative or qualitative data, or might be a problem of design. There may be an "obvious" answer to the problem, or it could be an open-ended exercise in creativity. Students are then arranged in groups of three (triads), and a specific role is assigned to each member of the group.

1. **The problem-solver.** The task of the problem-solver is to try to solve the genetics problem while speaking about the logical steps and thought processes he or she is going through. He or she begins by reading the problem out loud, and then proceeds with piecing the solution together using either linear or lateral thinking, as appropriate to the problem.

2. **The responder.** The task of this individual is to make sure the problem-solver clearly identifies the steps he or she is following in a logical manner, and explains his or her thinking. The responder then asks questions to clarify what is being said and why. The types of questions asked would include: "Why did you say that?" and "How can you conclude that?" and "What principle are you applying here?" The responder can even prompt the problem-solver if he or she perceives the person to be going off track (e.g., "Aren't we supposed to be finding out so-and-so?"). The responder is to avoid taking over and moving into the role of the problem-solver and he or she should allow the problem-solver to make mistakes and get stuck. Any difficulties that emerge can be recorded and may serve as the basis for discussion later (see **scribe** below).

3. **The scribe.** The job of this person is to record the sequence of transactions between the problem-solver and the responder. In particular he or she is to note successful avenues of attack on the problem, and, just as important, the dead ends and less effective pursuits. At the end of the problem exercise, the scribe reports the group's results to rest of class.

experiences to the rest of the class. In the ensuing discussion, he or she then compares and contrasts methods with other groups' scribes.

This triad method of problem solving has great potential in familiarizing students with the thought processes inherent in any subject. The most common difficulty we have encountered in teaching genetics problem solving is that many students individually cannot get "off the ground" when given a new problem to solve. Even after they have been presented with and studied the necessary principles for solution, their understanding of these principles is often vague, and their ability to apply the concepts weak. Verbalizing their thinking with peers, in groups, is a useful way to solidify ideas, reveal weaknesses and gaps in knowledge and understanding, and establish efficient patterns of analytical procedure. Furthermore, listening and speaking with peers who are non-experts is less intimidating for students than presenting ideas before an entire class. We have observed that an essential aspect of implementing this (and any) learner-centered, problem-solving strategy involves convincing both instructors and their students that this is a valid and useful method. Many professors and graduate student instructors have more experience and are more secure using a teacher-centered, "modeling" type of approach (e.g., "This is how to do this kind of a problem.") and are hesitant to try a new strategy even when modeling problem solving doesn't work. Students typically have had little experience in verbalizing intellectually, and most of their prior learning experiences have involved assimilation and memorization. However, both of these patterns of behavior lead to failure in problem-solving courses. Moving away from deeply held traditional views regarding the roles of teacher and learner and attempting this alternative approach is challenging for all parties, but worth the effort.

We encourage you to give the triad method a try in your classes, and modify it according to your specific genetics teaching situation.

References

Pestel, B.C. (1993). Teaching problem solving without modeling through "thinking aloud pair problem solving." *Science Education.* 77(1), 83–94.

Whimbey, A. (1984). The key to higher order thinking is precise processing. *Educational Leadership*, 42(1), 66–70.

4

Understanding the Understanding of Genetics

A common complaint of genetics students goes something like this:

"I *understand* all this material (on genes, chromosomes, DNA, etc.), but I still do poorly on exams, and consequently I am about to fail this course."

The frustration of such students is that although they believe they understand the genetic principles and analyses in the curriculum, their understanding is not being given its due credit in the form of assessment marks. Do these students have a point? Do they really understand? And how should instructors fairly assess student understanding of genetics? The notion of understanding is centrally important: Most students seem to prefer to understand rather than just acquire superficial knowledge, and of course instructors want to structure their courses to promote student understanding. Therefore understanding is clearly one of the main goals of any educational process. Thus, it is worthwhile reflecting upon the nature of understanding, how it is acquired and assessed, and the high expectations we have of it.

At the 1996 meeting of the AERA (American Educational Research Association) in New York City, an interesting session took place that shed light on the nature of understanding in science generally, and in genetics specifically. The session was a panel of experts discussing how best to define the elusive construct of understanding. Each speaker presented their view of understanding, which emanated from empirical studies they had conducted. The first study described was one in which students at the University of Edinburgh who were studying for their finals were interviewed about the methods that they were using to construct their understanding of the various subjects in their specialties. The British system of final examinations is much more of an integrative assessment than any comparable university or college exams in North America. Students are expected to be able to interrelate just about everything they have learned in their university careers in one subject area, and even in different subject areas. Hence students in these institutions are good subjects for asking how one can come to 'understand' a discipline such as genetics. The interesting result was that most students had devised a representational image of their subject area. This took the form of a distinct and integrated body of knowledge that could be vividly recalled in almost a sensory way. Noel Entwistle, the researcher who supervised the study, referred to this visualization as a "knowledge object" (Entwistle, 1996). Different students devised quite different types of knowledge objects. For one student the knowledge object was like an octopus with many branched and interconnected tentacles representing the various interconnected ideas of the subject. For another, the knowledge object took the form of the flight of a bird across the landscape of the subject, with all its fields and roads laid out below. Indeed many of the representations were map-like, calling to mind the idea of the concept map, which we describe in Chapter Ten. However, although the particular form of the knowledge object was heterogeneous, the commonality was the vividness of the representation—almost a living thing in its own right. It seems then that the student mind (at least in Scotland) has a penchant for generating knowledge objects, and success in the final examinations seemed to depend on the skill and creativity by which these representational knowledge objects are constructed.

In genetics one can imagine a range of knowledge objects covering various views of the subject. There could be a knowledge object for the entire discipline of genetics, a broad perspective sweep showing the general relationships between the subdisciplines—the "chapters" of the book of genetics. One can also imagine knowledge objects geared to a specific recurrent theme such as recombination, with tentacles ramifying into cytogenetics, molecular biology, evolution, as well as classical genetics. However, knowledge objects seem to be slithery creatures that can ooze to new positions and engulf ever-changing topics. Furthermore when one knowledge object encounters another, either they can fuse to form one bigger holistic vision or remain separate and distinct, depending on the specific learning need involved. Also there are undoubtedly organismal knowledge objects out there roaming the examination halls: one monstrous and fruit fly-like, another distinctly yeasty-looking, another vermiform, and so on.

While we sat and listened to the experts debating the value of this notion of understanding, the knowledge object beast was attacked (and in our view slain) by a Saint-George-taking-the-form-of-Harvard education researcher, David Perkins. To Perkins, representation, however useful, is not equivalent to understanding. Perkins argues that no matter what we do as educators or psychologists, we can never *really see* what goes on in the minds of students (Perkins, 1996). According to his logic the intellectual bouillabaisse of the mind is undoubtedly very important, but understanding can only be recognized and thus defined when certain types of behavior emerge out of this mental soup. Based on this viewpoint, individual knowledge objects are probably rather like undergarments—important and fundamental to the owner, but highly personal and not useful to others.

Thus, Perkins claims the key to understanding is performance. In fact he goes several steps further to maintain that performance *is* understanding. He defines understanding boldly as "flexible performance capacity." Although the idea that students need to perform during exams is not exactly new, the new twist is that the performance is the only reality that we, as educators can give to understanding. Understanding does not exist in any other real sense. All else is bouillabaisse. Flexibility is of course crucial to this viewpoint: given the complexity of a subject such as genetics, the student actor has to be able to process and rearrange the concepts and data in a potentially limitless number of ways in order to give a satisfactory performance.

This indeed is a concept that educators can sink their teeth into, an idea of immense usefulness in that it can not only shape but also evaluate educational practice. It points very strongly to the value of student-centered learning methods we discuss throughout this book. Perkins's concept of understanding implies that making notes and reading books are just the beginning of developing understanding, just a reading of the script prior to the real thing, which is the actual performance. And, of course, as every actor knows you simply cannot put on a decent performance without rehearsals; the more of them the better. Understanding simply does not snap together during exams, it arises out of repeated attempts to flex the performance muscle.

If we accept that understanding (of any domain) *is* the capacity to perform creatively, we need to consider what we mean by performance. That is, what forms can performance take? Obviously the performance can be spoken, and indeed our experience with verbalization activities and oral exams is that they very effectively reveal the strengths and weaknesses of student understanding. Possibly a spoken performance is the most difficult yet the most desirable performance mode. Which professional geneticist has not discovered the rule that "You never understand a principle or concept until you teach it"? And, reflect back on all the M.Sc. and Ph.D. defenses you have attended! Many of the students in graduate orals have had the opportunity to perform in this setting, and it shows. Despite the attractiveness and value of using spoken performance as a measure of understanding, there are some difficulties associated with this format. Caution is needed in comparing different personalities, temperaments and social backgrounds. Cultural attributes and foreign language difficulties are also issues. Furthermore this type of assessment is not practical for use in the megacourses, which are commonplace in many institutions.

Written tasks are also forms of performance, and are probably more universally applicable. These can take numerous forms during class-time or, alternatively, in test situations. If the written task is taking place in an exam setting, it is important that the questions asked give full opportunity for the learner to demonstrate performance skills. Questions of the type "How are genes transcribed?" are not likely to lead to creative performances. The more connections required the better.

Because of its complexity and intellectual richness, genetics is ideally suited for acquiring the flexibility needed for a convincing performance, be it written, spoken, or other.

(Whereas we have not considered the use of drawing or dramatic role-play here, we have successfully used such performance modes to explore student understanding in genetics.) Consider the multitude of different layers of the subject, ranging through the molecular, chromosomal, cellular, tissue, developmental, organismal, populational, and evolutionary levels. Consider also the different life cycles, the different model organisms, and the language, all of which need skillful integration for a convincingly flexible performance.

Genetics is traditionally assessed by asking students to solve problems. The successful solution to a problem is clearly a type of performance, so the traditional assessment practice of genetics is a sound application of the Perkins principle. However, the exercises that lead up to the final assessment performance in problem-solving might also be performance-based but of quite different types (e.g., any type of data processing by students, either individually or in group activities in class or tutorials.) In this book we advocate the use of such exercises as think-aloud pair problem solving, concept maps, problem expansions and multiple representation exercises as methods to promote performance skills. Unfortunately, left to themselves, students who are firmly adapted to passive learning modes will not voluntarily enter into these performance rehearsal exercises. These students will need guidance and encouragement, and an explicit discussion of what is to be gained by engaging in nontraditional, creative performance activities.

It goes without saying that the Perkins definition throws into high relief the idea that assessments based on memorization and recall have little to do with understanding. Although the act of recognizing a word on the exam paper and blurting out a remembered set of words is a kind of performance, it is a low-grade and inflexible "Bottom-and-Quince" type of performance.

Performing skills are in great demand in the world outside of the ivory tower of academia. Interpersonal communication, public speaking, teamwork, management, brainstorming, and report writing are all skills that are heavily drawn on in business and

31

government, and all represent various forms of performance that demand high levels of integration of ideas and a nimbleness of expression. In fact, according to several statements issued by business CEOs, these types of generic skills are those valued most highly when these companies seek new employees. In other words, an individual's subject-matter knowledge is secondary to skilled problem solving and versatile delivery. Therefore it is likely that the virtues of developing generic skills in performance-level understanding range far beyond the genetics course into the realm of careers. So next time students complain about the excessive rigors of learning genetics, and about the fact that they really understand an awful lot more than they have been given credit for, remind them of the Perkins definition. Then, point out that you are developing in your students a competence that will prepare them for life-long learning in the "real world."

References

Entwistle, N. (1996). *Knowledge objects as frameworks for understanding and explaining*. Paper presented at the annual general meeting of the American Educational Research Association, New York.

Perkins, D. (1996). *Image as insight: The role of representations in understanding*. Paper presented at the annual general meeting of the American Educational Research Association, New York.

5

Geneticspeak:
The Role of Story and Language in Teaching and Learning Genetics

Language and storytelling have been used by humans for thousands of years as a way of preserving history and culture. Elders telling stories to groups of people sitting around the fire on winter nights have probably transmitted more bits of information over time than all the world's libraries put together. On the West Coast of Canada the traditional native tales of the raven, the thunderbird, and the wild woman of the woods are good examples of the way in which the richness of language is easily retained and transmitted. And of course along with the language (as the vector) go the morals and traditions that constitute the culture. So the story is not all entertainment; it carries along with it a considerable measure of social code of behavior, which is long remembered.

What role do language and storytelling play in science? In what form does one store scientific knowledge in the brain? Is it in fragments like mental stone tablets representing a number of current but discrete truths? Or is it in the form of some type of smoothly flowing story? One way to think about this is to wonder if someone were to ask you what contributions you have made to science, how would you respond? The chances are that you would launch into some kind of narrative about why you started on this or that type of research, what was known at the time you began, and then how your experiments caused the scales to fall from your eyes. In other words, it seems that for many people information about science is stored as some kind of story, and retrieval is like a fast rerun of the story "tape."

How does this relate to the learning and teaching of science, and more specifically, genetics? If we as professionals, with our wide exposure to scientific facts and our rich supply of scientific experiences, routinely store and access our knowledge via a series of interconnected stories, should we assume that students who are just beginning to wrestle with the subject of genetics do the same? In what form is information stored by these novice practitioners? While in theory it could be the same as for professionals, it appears that this is often not the case. For many students, information, concepts, and interrelated principles appear to be stored as distinct modules or units. How often have we observed that students cannot connect the different aspects of genetics, or that they seem to trot out a learned module or slogan whether or not it accurately fits the question under discussion? This observation is raised not to allocate blame, but rather to promote a search for solutions.

If stories provide an important and useful system for the storage and retrieval of information, it seems appropriate that scientific information should be acquired in story form. Professional scientists acquire most of their information through research experiences or through the experience of reading primary research articles, both of which have strong storylike qualities. How do students acquire scientific information? Regrettably, most likely by reading about or listening to lectures providing a teeming myriad of facts or a plethora of principles. Consider the modern textbooks in virtually any area of the life sciences; most are at least 1000 pages long. Much of the text is chopped up into short, discrete, "readable" subsections, designed to provide students with reasonably small and easily digestible bites of information. This design is well intended but conveys the message that learning science is all about rote mastering of a huge body of discrete, and often unrelated facts and principles. With this as their model how can students be expected to input that information in an integrated way that resembles a story? Yet this is precisely how we expect students to perform when we examine their understanding or otherwise assess their progress. We expect them to assemble these components into an integrated story. In fact the word story is often used for this synthesis.

Language itself is the other component of the formula. Of course language is necessary to tell the story. One of the complaints we commonly hear from students about genetics is that there seems to be an excess of specialized terminology peculiar to the genetic language, what we could call (with apologies to the ghost of George Orwell) "geneticspeak." It appears to students that geneticists have made up words deliberately to confuse them, or to keep genetics as insider information. Yet the truth is that *the language* actually *is the subject*. All those difficult-sounding words, heterozygote, translocation, transcription, transposition, merodiploid, nondisjunction and so on, actually constitute the subject matter of genetics. The subject matter does not exist outside of its language. This is not a play on words; genetics stripped of its language would be so imprecise and unrecognizable as to make it no longer genetics. So, one could say that learning the language of genetics is learning the subject. It follows, then, that demonstrating an intimate familiarity with and ability to use geneticspeak is a good indication that one understands the subject. To learn and use the language of genetics is to become a geneticist, and to think like a geneticist. There is no avoiding this if genetics is to be mastered with deep understanding.

However, things get muddled in a classroom setting if *rote* learning of genetics vocabulary and definitions become the focus, without any associated deep understanding of concepts. The language must make sense right from the word go. One symptom of this problem is the student who can define a given genetic term, but cannot explain the connection between the underlying concept and related concepts. This is like learning and reciting a poem in Swahili! Making a concept map is a simple way of checking if students are stuck in this mode (see Chapter Ten). The difficulty carries over into problem-solving and becomes more serious because students are unable to explain how concepts apply to the particular genetics problem they are trying to solve. Whereas there is no simple way to convince students that superficial parroting of definitions is not the goal, getting them to practice speaking and using the terminology may help convince them that their understanding is not as deep as they might have thought. We suggest a few strategies for doing this later.

Another part of the problem in becoming proficient with scientific language is that many of the words of science are also used in everyday speech, but these words have quite different meanings in common parlance. A colleague teaching first year biology had spent the teaching semester talking about "animals" only to discover that for many students the term meant mammal. Somehow all the discussions of gills, tracheae, etc. had been force-fitted onto the concept of mammal! Is there a similar problem with the language used in genetics education? Sadly, the answer to this question is "Yes." In Chapter Seven we rail against the multiple uses of the word "cross" in the teaching of genetics. It is not difficult to locate many other examples. For example the word "strand" in common speech means one of the subunits of a bundle of threads. In genetics the meaning is similar, but the definition of what constitutes a strand can change with the discussion. In discussions of crossing over, a strand is a *chromosome* (as in "four-strand double crossover"). Yet in discussing DNA a strand is a *polynucleotide chain* ("DNA is two polynucleotide strands intertwined as a double helix"). Other examples used in common parlance include the terms induce, repress, enhance, translate, regulate, imprint and transition. The meanings in common speech are part of the baggage that learners bring to our genetics courses; the meanings cannot be expunged, they exist parallel with the specialized genetic meaning. Recognition of such dual definitions and practice with the language of genetics is the only recourse.

It is easy to overlook other sources of communication confusion that lurk in what appear to be the straightforward, everyday language of geneticists. Consider the geneticist's favorite shortcut for communication, the use of symbols. To the instructor, symbols are helpful, benign devices. Students experience them differently, however. Genetics contains excruciating cases of confusing allele symbolism. One that springs to mind is that in bacterial genetics the symbol lac^+, means "can utilize lactose," whereas leu^+ means "can synthesize its own leucine!" Therefore these nearly identical symbolic representations have quite different meaning in almost adjacent areas of genetics. Furthermore, whereas it might be true that it would be a better world (for students, in particular) if the symbolism used in bacteria, yeast, Neurospora, worms, flies, and humans was identical, an Esperanto version of genetic symbolism hasn't been devised, and the bewildering heterogeneity remains an integral part of genetics. Indeed the different sets of terminology contribute to the richness of the subject in the same way that

LINES

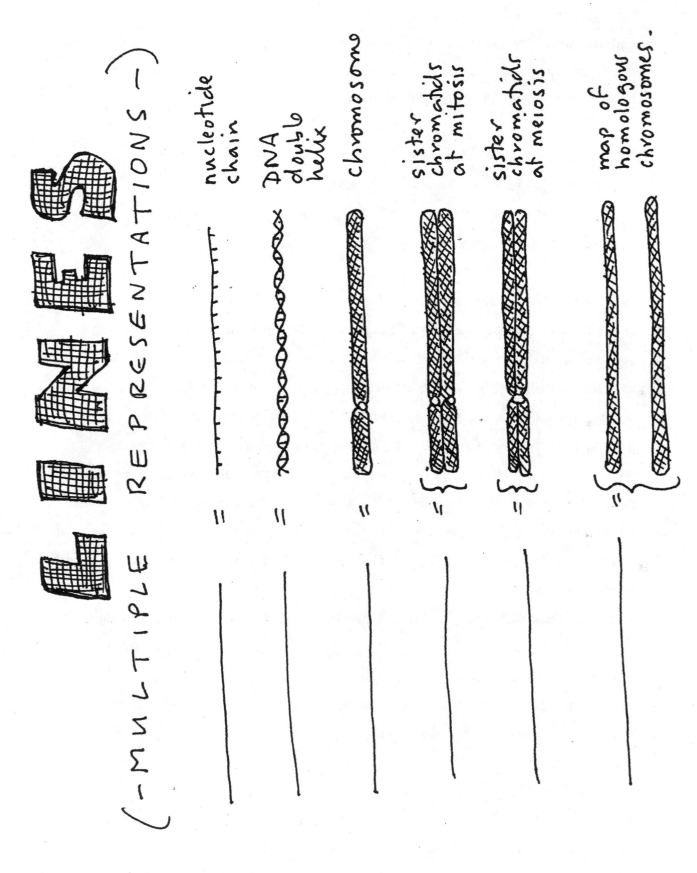

nucleotide chain

DNA double helix

chromosome

sister chromatids at mitosis

sister chromatids at meiosis

map of homologous chromosomes.

the diversity of human languages contribute to the richness of the human condition. Unfortunately they also contribute to learning problems in our students. Thus, before the rich tapestry of genetics can be understood, these multiple representations must be learned, and this can be enhanced through the use of stories.

We have used a number of strategies to assist genetics students to develop an understanding of genetics language and put their own set of stories in place. Some suggestions for you to try are listed below. (For other ideas on how to explore genetics language use with students see Chapter Six.)

1. Use stories in lectures

If stories help organize knowledge for easier retrieval, we can assist students in learning genetics by incorporating stories into lectures. As with stories told in native cultures, or the parables of Jesus, the listeners are personally drawn into the story, and internalize the content without consciously having to work at it. One excellent source of stories for genetics instructors are the "Perspectives" articles in the journal *Genetics*, in which the authors reflect upon the history or development of some particular genetic advance or some great geneticist they have known. Undergraduates at our university seem to appreciate hearing about how, as a young undergraduate, Sturtevant devised the theory of chromosome mapping in one evening. Other stories we use seem to be more apocryphal. Another story we have used is how Sparrow inferred the mutagenic effects of chemicals by inferring that plant mutations in the Brookhaven greenhouse were caused by the proximity of the air intake duct to the chemistry department fume hood exhaust on the roof of the building. Such stories may have become distorted in the telling over the years, but still have the virtue of holding the students' attention, and conveying a train of important concepts linked together by the form of the language.

2. Student "quickie" presentations

Without realizing it, many students read their textbook and notes in a superficial manner. Consequently they fail to recognize and note the crucial elements of a topic. Even students that try to take a deeper approach to studying have difficulty putting what they read into words. One solution is to encourage students to actually verbalize what they think they have learned, either through talking to themselves or to a friend. However, the success of this approach depends on the personal initiative of students on their own time, and this cannot be relied on. A more formal and structured way to give students practice speaking about genetics is through short presentations to the class. The students in a class all spend some preparatory time researching various assigned topics or concepts using the textbook or other sources, and come to class prepared to give a short presentation on one or more topics without notes. To encourage students to attend to their peers and to provide additional practice with presentations, these strategies can be combined with the activity we call Models and Mimics in Chapter Six.

3. Interpretive talk about figures

Diagrams, graphs, and photographs in texts are often overlooked by students when studying and reading. Furthermore while these various types of text figures are designed to enhance written text and convey additional information, our experiences show that what the authors intend to illustrate is often ignored or misconstrued by students. For these reasons it is valuable to use text figures as a source of interpretive speaking activities. This procedure involves asking a student to give a two-minute talk about some relevant figure in the textbook, say, a diagram or a photograph of an interesting phenotype. For example, students may be asked to interpret what is being represented in the text figure directly, or they may discuss how a given diagram can be connected to a particular genetics concept or principle. These types of tasks give students practice with the language of genetics, while assisting them in recognizing there is a relationship between "static" text figures and the dynamic biological processes they represent.

4. The one minute monologue

We have found that genetics students become severely tongue-tied when asked to speak concisely about genetics principles. This is usually manifest as an awkward silence when students are asked to describe a concept in their own words. To provide them with some skills in this area, we devised what we call the "one minute monologue." This is based on the old BBC radio show called "Got a minute?" The strategy involves having a student draw a topic from a hat and then speaking for one minute on the topic without notes. The goal is that their monologue must be free of repetition, delays, and hesitations. The topics should be reasonably challenging but not too big, such as "the diagnostic tests for gene linkage." It is important to introduce this strategy with concepts and ideas that students are very familiar and comfortable with and progress in complexity as students become more adept. Positive feedback for contributing is an essential element of this strategy, as students initially find this extemporaneous speaking very risky and stressful.

5. Vocabulary practice

There are many ways to practice with the vocabulary and symbols of genetics. In an introductory course ask students to comment on symbols with similar meaning such as $+$, a^+, wild type, and A. Or ask them to compare and contrast different representations of the same basic process such as pedigrees, Punnett squares, and branch diagrams. Ask them to make up sentences using three related genetic terms. In advanced courses we all assign papers for students to read and then hold classroom discussions which necessitate speaking. But most students we encounter need basic practice using the language of genetics before they take on the job of reading and discussing a scientific paper.

We have argued above that the language of science is an essential aspect of scientific discipline. Yet, too often in the classroom the only people actually using language are the instructors. Many students believe it is their lot in life to listen and our lot is to speak. This convention needs to be challenged and turned on its head. As most of us have discovered, it is only when we teach (and speak) about something that any underlying confusion becomes evident and our understanding becomes clarified. Speaking a language provides a learning experience besides which all bookish exercises of copying out verbs and declensions pale. In order to learn genetics, students must become proficient at "geneticspeak" because only then can enter the professional community of science.

6

Models and Mimics

Most university and college teaching involves some form of modeling. A professor stands in front of a class and shows students how to reason logically in their particular subject area. Generally, genetics professors do this rather well. Drawing upon their expert knowledge, experience, and insight, they manage to put on a good show of modeling how an experienced professional in their discipline thinks about the subject matter. Teaching assistants, acting in their capacity as tutorial section leaders, also model. Emulating the professorial style (they are, after all, professors-in-training) they demonstrate how to solve a problem selected from the assigned homework, or they expand upon or clarify some topic from the course. In part, modeling involves the direct transmission of facts and information, but in many cases it also involves the sharing of process. Consider, for example, the conscientious professor who through modeling is attempting to communicate to students the ways in which inference is made in his or her subject. Regardless of whether the focus is fact or process, the student's job in the educational settings just described, is simply to imbibe and assimilate the presentation, and then later pass an exam using the same type of logic. On the face of it the formula seems straightforward and unproblematic.

How do the students feel about this? They appear to love and cherish modeling. They love to hear the answers to the challenging scientific questions posed by their teachers, and they wait with baited breath to hear the answers to homework problems, all expounded elegantly and clearly. Indeed, they reward such teachers with high scores on departmental teaching evaluations. But how do these favorable feelings translate into deep understanding and mastery of the subject? There is no doubt that having a clear professor as a model is much better than having a muddy thinker or a disorganized lecturer. However, as we have suggested previously in this book, being clear is simply not enough. The velvet tongue and inspiring performance of the lecturer, the logical and clearly flowing set of notes transcribed, can all act to lull the students into thinking that they, like the professor that stands before them, have mastered the concepts. In the students' view, a complete understanding of the subject resides in their set of notes. Just having their notebooks full gives them a sense of security and power. Unfortunately in many cases the ideas on the notebook pages simply do not seem to get transferred into the students' brains. At exam time, instead of the powerful analytical approach of their professorial model appearing on the exam page, answers emerge as a series of fragmented "sound bites" and half-truths dimly remembered, drawn partly from perused notes and partly from deeply entrenched scientific slogans learned in previous school experiences.

Making a change in this situation can be challenging. In this book we discuss a variety of exercises we have designed and used to promote active processing of information by students. We have discovered however, that there are both teachers and students who, when introduced to these exercises, experience some real problems "buying into" the notion that such activities are beneficial in making progress toward the learning goals. Many students seem to feel that they are being short-changed by the new methods. Because for most students their previous exposure to education has been a system whereby passive information acquisition and recall have been rewarded, when they are confronted with new approaches that require active processing, risk taking, public activity, and potential for embarrassment, they balk. We have found that a common response by students to anything they regard as unorthodox instructional practice is to direct the instructor to get back to the "real" teaching of science (that is, to get on with "providing the facts"). Indeed, recently a colleague who was attempting such approaches was scolded by a student who made the following comment "I've paid a lot of money to come here and be taught by an expert, and that is what I want. I don't want to talk to my fellow students about their ideas" (a reference to some of the processing exercises). In truth there is often no simple way of convincing a student that engaging in processing exercises will pay off in terms of higher grades—for one thing, in learning there generally is no "controlled experiment." Even with the best of successes in new processing strategies it is difficult to overcome the skepticism wrought by a lifelong learning habit in just one course.

Our discussions and work with faculty and teaching assistants indicate that they also feel uncomfortable when administering non-traditional instructional approaches that require the students to process information (e.g., Think Aloud Problem Solving in Triads [TAPS] and Concept Mapping, which are described Chapters Three and Ten respectively). Using these types of educational approaches requires that the instructor possess a broad and deep grasp of the subject material, and enough poise and dogged determination not to get flustered and to stay on track. The usual rewards of teaching—good evaluations, a feeling of self-worth for serving as a source of wisdom for students, a feeling of being an expert, a feeling that the class is with you—might all fly out of the window.

Why try to change the status quo if everybody seems to be happy and no one seems to be listening? Because we argue that traditional instructional approaches are both unsuitable and inadequate, and they contribute to a continued focus on memorization and a lack of engagement with the type of data analysis that constitutes the heart of science as practiced. During some of our more cynical periods of reflection we fret about the image of science that has been generated during the university education of science graduates. The model of the scientist lodged in these young minds appears to be that of a walking encyclopedia of science, and that is what many of these young people have become. Somewhere along the way our science education system is not working the way we think it is. We believe it is important to provide a more authentic view of science and the scientist, and thus continue to promote understanding and acceptance of active student-based learning.

So where can one start? In what follows we present an educational strategy that we have found to be the most successful in overcoming some of the skepticism, reluctance, and resistance of instructors and students to exploring new educational procedures. The method requires some minimal risk-taking by all involved, but the level and form of this risk-taking seems to be acceptable to all participants. The method takes advantage of the penchant of all parties for modeling. Instructors like to model, and students like to see modeling, so why not take advantage of this situation and try to harness this unquestioning faith in modeling to promote thinking about the importance of active participation?

We call the approach "Models and Mimics." The professor or teaching assistant acts as the model, and, as per tradition, they illustrate a task or problem by modeling while the students watch. Then students become the mimics; they must *exactly* duplicate what was modeled. The principle behind this method is the fact that mimicry is not a low-grade mental activity. For mimicry to work, the ideas and facts of the modeled system must be accurately and actively processed in the learner's brain in order for them to be able to walk through the identical analytical procedure explained by the modeler. For communication of the process to take place, either verbally or in written form, all the mental circuits have to be primed.

This activity illustrates some important pedagogical principles. First, learning by simply watching an expert who is modeling something is exceedingly difficult if not impossible. Second, this type of learning is not only hard work, it also assumes that the learner has in place most, if not all of the prior knowledge of the expert. Third, complete mimicry goes well beyond duplicating the behaviors of the expert. To replicate the thinking of the expert, the mimic must have acquired a complete understanding of the concepts being modeled. These principles do become apparent to most students, after they try to accurately mimic their instructor. However, to guarantee that the point of the exercise is completely understood by all participants, it is very important to hold a discussion focussing on *why* most students cannot perform perfect mimicry of a modeled problem.

Incidentally, we find that it is true generally that students do not mind the instructor giving a name to processing exercises, such as Models and Mimics. If a justification for using the procedure has been well explained, and the method is introduced slowly and with confidence, then the new instructional procedure can become part of the accepted routine and not a new weird diversion that students believe will distract them from their "real" learning (i.e., memorization).

We have used three variations on the Models and Mimics theme. To bring out the important pedagogical principles discussed earlier in the chapter and to gain familiarity with the entire process, we recommend that Variation 1 be mastered before moving on to Variations 2 and 3.

Variation 1: Pass the chalk

Pass the chalk is a procedure that lowers the risk of students presenting in front of their peers. The basic premise is that when a student who is presenting his or her ideas (i.e., the mimic) feels unsure or uncomfortable, he or she simply passes the responsibility (in the form of chalk, whiteboard marker, or overhead pen) to another speaker who takes the stage.

To begin this procedure the instructor selects a new problem for modeling before the class (one never seen or solved previously by the group) and then explains the approach and the "rules." The professor tells the class the following.

> I will model the method that I used to solve this problem, including the diagnostics that were helpful in moving stepwise from the data given to the solution. It is not the only possible interpretation of the given data but it travels a route that many geneticists would use, so it is worth analyzing. For this reason I ask that each of you try to reconstruct this pathway. Therefore please listen carefully and make notes. When I am finished I will give you several minutes to collect your thoughts and then ask one of you to mimic my presentation. To do this you will repeat my solution to the class without using your notes, reconstructing *all* of the intermediate steps.

The instructor then explains that no one student will be held responsible for doing this entire task. Everyone must attend to what is happening because if the selected student gets "stuck" somewhere along the way they can pass the chalk to any other member of the class who will then take over as far as possible and then pass the chalk again. Although the description refers to modeling and mimicking a new problem, the process can be applied to any type of explanation, quantitative or qualitative, or a problem all students have seen and found challenging.

Unless the solution involves complex diagrams, any one pass through this activity usually takes less than five minutes and involves between one and three students (although there are no limits to how many can participate). The instructor facilitates the mimic but doesn't interfere or intervene unless the student is really stuck. In such a case the instructor can ask a question to assist the student in examining the logic, sequence, and clarity of their own presentation. Other students should be asked to also participate by providing this type of support. The student should be made to feel comfortable and allowed to leave " the podium" whenever he or she wishes and another student will take over.

A common outcome is that students who think they "know" the solution get bogged down because of incomplete or shallow understanding. Even the students who seem less challenged by the problem have an opportunity to practice expressing themselves verbally, and often come adrift as a result of some poorly understood or poorly worded idea. One run of pass-the-chalk is not necessarily the end of the process. There is ample reason for asking for a new voyage through the same problem because this will involve more students and should bring out more streamlined explanations. The possibility of a second pass keeps everybody on his or her toes. For the next modeled problem, a new set of students is asked to participate. The final and most important phase in the activity is a debriefing of what was learned about teaching, learning, and knowledge of the concepts. This discussion should elucidate problems associated with learning by listening and watching an expert model. A parallel discussion should focus on what students can learn about their grasp of concepts when they mimic an expert.

Variation 2: Making overheads

This variation should be introduced after your class is aware of the issues and limitations of learning from listening and watching. The basic procedure discussed above still holds for this variation of Models and Mimics. In this case, however, the class is divided into groups of four or five and each group is asked to prepare a solution to a previously modeled problem or other presentation (e.g., "The genetic criteria used to distinguish an inversion from a translocation"). After about ten minutes (don't let it drag) the group writes their report (this is part of the "mimic" process) on an overhead sheet for presentation to the rest of the class. Ideally, the instructor nominates a member from each group to start explaining the solution with the aid of the overhead sheet. The mimic in the first group thus also becomes the modeler for any other groups that follow. Alternatively the instructor leads a general class discussion. The instructor will display the students' solutions and ask the class to work as group, identifying difficulties with wording, and commonalities, and differences between the solutions.

One of the valuable spin-offs of this type of exercise is that students in the class get to know each other. In university situations where there are very high enrollment classes this approach provides a humanizing effect. By moving away from the strict instructor modeling this teaching approach allows small encounter groups such as tutorials to promote student socialization. Indirectly this promotes a better learning environment because the students will feel more secure speaking in front of their peers.

Variation 3: Board work

Most students have an understandable reluctance to go the board to explain something. This version of models and mimics is similar to the Variation 2 except that a long room-sized chalk or whiteboard (assuming one is available) is divided up into four or five areas representing different assigned problems (from homework or unseen), and a group of students works together at each station. They can use notes or the textbook, but eventually their board presentation becomes the basis for their report to the rest of the class. In this approach the reporting student is not necessarily mimicking some specific piece of modeling performed by the instructor, but the general presentation style used in the subject of genetics.

The board method has also been used successfully as a review session just before an exam. The divisions on the board then become the main topic areas covered for this particular exam, whether midterm or final. A student group assembles the main principles and diagnostics for their assigned topic area, and reports on these to the rest of the class. The feedback we received from students on this method has been very encouraging; students claim actually to see light-bulbs going on during this active process, and there is the added benefit of seeing the entire curriculum and its main points laid out like a map. Just as in the case of the overhead method, there is an active and focused social buzz that illustrates processing is occurring.

In summary, what are some of the benefits from trying Models and Mimics?

• First, the method can provide important insights for students. For example, most students (and instructors) are surprised that the seemingly simple task of mimicking the instructor is in fact exceedingly difficult. This may serve to shake up students that have become complacent and passive during lectures. It also helps students realize they understand less than they had thought about a particular concept or principle. Students appreciate discovering their area of confusion early in the course rather than during an examination.

• Second, this approach provides students with different types of practice with the course concepts, that is, creative performance opportunities (see Chapter Four). Students have an opportunity to explore using discourse in science, the vocabulary of genetics, as well as practice with the particular principles presented by the modeler.

• Third, this procedure provides opportunities for students to socialize and communicate with peers. Both are significant and valuable components of the university educational experience.

• Finally, although Models and Mimics is a very modest switch from traditional practice, it can serve as a transitional stepping stone to tempt instructors and students to try out more adventurous new strategies.

7

Too Many X's
or
The Cross Genetics Students Have to Bear

In 1998, the organizers of the Toronto International Genetics Congress ran a competition for a congress logo. At that time we spent some time thinking about what image best represents genetics, and concluded that the cross "X" symbol comes the closest. Sure enough, the winning entry was a stylized cross. However, the cross is a much-overworked image, not only in genetics but in life generally, and this is a source of great confusion to students trying to untangle the hieroglyphics of genetics. Professionals are easily able to flip from one meaning to another but this is tough going for a novice. In this chapter we consider some parts of the conceptual minefield that this plethora of crosses presents to the learner.

First, let's consider some of the cross symbolism "baggage" that the student brings to the course from everyday usage. What does a cross conjure up to the average person? A few diverse examples are (in no particular order) Christianity, a railroad crossing, the signature of an unschooled individual, death, a cartoon representation of the eyes of a drunk, the letter X, X-rays, and, of course, kisses. Furthermore, in mathematics X can represent an unknown variable in an equation or a multiplication sign.

With these multiple everyday conception of X's buzzing in their heads like so many flies, the students enter the *terra incognita* of a genetics course. First they are told that a cross represents mating, which of course geneticists actually call "a cross." Probably the next way they encounter a cross symbol is when it is used to represent the X chromosome, usually without being told exactly why it is called an X chromosome. In addition, some texts and/or instructors actually draw an X to represent the X chromosome diagrammatically, with tiny alleles arranged on it like superscripts; this usage is a huge landmine that can explode if it is used in conjunction with the use of other very similar symbols described below.

The problems associated with multiple visual representations may be compounded by aural confusion. Phonetically the "ex" of the X chromosome sounds very similar to the "sex" in the phrase "sex chromosome," and these two similar sounding terms *may*

compete in the novice brain for significance. Other possible sources of phonetic fumbles include terms like exconjugant, extrachromosomal, exon, and exogenote—all of which contain the "ex" sound in quite different constructions. Although the role of "ex" in each of these terms is quite different, any one could be confused for "X" if these concepts are presented only orally during a lecture.

Students are next introduced to the concept of crossover, often represented in textbooks and lecture diagrams by (of course) a cross. Once again there is a confusing phonetic juxtaposition, this time of the words cross and crossover. And, as if the cross wasn't well and truly worn out by this time, it resurfaces in a form that is rotated 45 degrees, appearing as the vertical cross (or + sign) that represents wild type alleles. In its form as a wild type allele, the cross may assume any one of a number of positions on the written line to give us the interchangeable symbols $+$, a^+ and $+^a$. This allele symbolism becomes even more complex when compound constructs of crosses such as X^+ are used. Soon after this in many courses, the students are probably introduced to the use of the cross in designating hybrids, and then almost certainly to the cross-like symbol for chi-square.

However, we have saved the best (that is, the worst) for last. Probably the single most confusing image in teaching and learning genetics is the practice of representing chromosomes (generally autosomes) as their mitotic metaphase X-like appearance. If students could grasp that these are mitotic metaphase conformations, then all would be well. Unfortunately, this X representation of chromosomes is often learned early on, in high school or in previous courses, and is so universally accepted that the students automatically assume that this "X" is the normal appearance of *all* chromosomes regardless of their stage in the cell cycle.

The consequences of this naïve conception are horrendous, and lead to several serious learning problems, not the least of which is the inability to distinguish between mitosis and meiosis. Admittedly, even when students do understand that interphase chromosomes are better represented by single lines than by Xs, they still confuse mitosis and meiosis. However, the X representation exacerbates the problem mightily. Consider some pitfalls. If the student believes that chromosomes are like Xs, then the first bewilderment that confronts them is whether this X represents one chromosome or somehow represents a homologous pair. The next mix-up comes when chromosomes form chromatids. Now their X-shaped chromosomes take on a double-X appearance resembling a meiotic structure even though this might be mitosis that is being considered! If a student *is* in fact

MULTIPLE REPRESENTATIONS

PEDIGREE

PUNNETT SQ.

Aa × Aa

♂ A   a

	A	a
A	AA	Aa
a	Aa	aa

♀

BRANCH DIAG.

$\frac{1}{2}$A → $\frac{1}{2}$A

$\frac{1}{2}$a → $\frac{1}{2}$a

→ $\frac{1}{2}$A

→ $\frac{1}{2}$a

♀   ♂

trying to reconstruct meiosis, and knows that meiosis involves pairing, they may end up trying to pair such double-X structures, the result is an eight-stranded monster, which is actually often seen roaming the pages of exam books.

Even the legitimate use of an X for a pair of sister chromatids held together by the centromere at meiosis is subject to misinterpretation. A common error is to view the X as two crossed chromatids, rather than as two parallel chromatids pinched together at the centromere. The crossed chromatid view leads to major errors in handling crossovers and segregations (try this out).

The moral of this story is "Beware of multiple representations!" Genetics is full of such examples in addition to the cross symbol. Each representation is imbued with a large amount of meaning that is well-known and distinguishable by practitioners. These alternative meanings (whose differences are often exquisitely subtle) would probably still be difficult to understand even if they were represented with clearly distinct symbols, but when represented by the same symbol it is little wonder that students get cross-eyed (confused, that is).

So what can be done aside from developing a new system of nomenclature and symbols? One place to begin is by asking students to keep a list of the different representations of one specific genetic item such as the cross symbol and to update their lists as new representations are introduced in the course. This can operate as a consciousness raising activity and serve as a clarification task. In a more comprehensive procedure that could move students from simple recognition toward distinguishing the multiple representations, consider identifying the text locations (page numbers) of a number of similar but distinct X representations and asking your students to explain how each representation is similar to or different from the others. A related but more difficult task would be to ask your students to locate and identify multiple representation within a section of the text or a set of genetics problems. Indeed, any activity that focuses on developing recognition of and providing practice with these similar and overlapping representations will help students bear this burden of so many genetic crosses.

8

Meiosis: Who Needs It?

Those lucky genetics students on the distant planet Zork! On Zork, most organisms are diploid and genes work more or less like they do on Earth, but meiosis is much simpler. There is no premeiotic S phase of DNA replication, so there is no chromatid formation. The homologous chromosomes simply pair, engage in crossing over at the two-strand stage, and then segregate into two daughter cells. Therefore all the processes achieved on Earth are achieved on Zork but with only one meiotic division. On Zork the genetics students never have a problem with meiosis—in fact they consider it trivial and go on to more challenging stuff.

In contrast, their brother and sister students on Earth are easily bogged down by meiosis. And no wonder—it is not easy to make sense of why it should involve DNA doubling and two cell divisions. How do the students accommodate this unreasonable complexity? Unfortunately they are hampered by several disorienting teaching traditions such as telling students that there is only one reduction at meiosis (the first division), whereas any thinking student can see that in fact if there is a doubling of genetic material there must be two reductions, one at each division. For some unknown reason we only emphasize the reduction of centromeres! Many students grapple with meiosis and eventually attain mastery over it. However, we have found that some students simply thrust the topic aside and don't deal with it. This remarkable discovery has emerged from studies on genetics students conducted by Patrick Lai during his doctoral studies in Education at the University of British Columbia (Lai, 1996). Lai ran a meiosis questionnaire in the first week of the genetics course, and then followed up with post-course interviews with a sample of students showing a range of grades in the course. One of the surprising findings was that many of the students who had passed the course had only a very dim understanding of the basic goings-on at meiosis. Their confusion wasn't limited to the details of the various stages of prophase I etc., but rather, dealt with fundamental aspects of chromatid formation, pairing, and segregation.

Most professors believe that meiosis is central and essential to genetics. Thus, the fact that many students can do reasonably well in genetics courses with little comprehension of meiosis should come as a shock to these people. But perhaps it shouldn't be surprising when we consider another of the ideas held dear by geneticists, which is that Gregor Mendel, the father of genetics, was able to deduce the basic ground rules of transmission genetics without knowing anything about meiosis! Indeed geneticists take pride from the fact that many of the processes of genetics were worked out by purely genetic analysis before the cellular or molecular basis of these processes were discovered. It seems that today's students might be emulating Mendel more closely than we imagined. Indeed it is interesting to ponder just what one can get away with in the absence of understanding meiosis. It seems likely that students can be successful operating with a set of basic principles of genetics such as the following:

- Mendel's first law of segregation of alleles
- Mendel's second law of independent assortment
- Recombination of <50% indicates degree of gene linkage.
- The dominant phenotype is the one shown by the heterozygote.
- Mutations often result in null function.
- The one gene-one polypeptide principle

None of these principles require a detailed understanding of meiosis, but this "kit" is probably enough to pass a basic genetics course if the concepts contained therein can be applied reasonably well. Even topics that would seem to need meiotic details, such as chromosome rearrangements and tetrad analysis, have their own extensions of the above kit of principles that enable students to process data in exams without deep understanding. An example of formulaic solution is that abnormally low recombinant frequencies suggest the presence of an inversion heterozygote; if this rule can be remembered and applied judiciously then the student will probably be able to press enough of the right buttons in the mind of an exam marker to obtain a passing grade. In genetics we pride ourselves that deep understanding is necessary in order to perform well in courses and on examinations. That is probably true, but on the other hand a machine-like approach to analysis probably suffices in many marginal cases.

Let's have a look at some more of Lai's findings.

1. The most common error he encountered was failure to recognize that there are two types of segregations of independent allele pairs (i.e., A B$\leftrightarrow$a b and A b$\leftrightarrow$a B). Most students show the first type only. This naïve conception of segregation haunted many students even into the post-exam interviews.

2. Students believed that at meiosis chromosome replication occurs but no pairing. This conception of meiotic activity reduced the first division to a mitotic division, and led students to generate some interesting schemes for the second division. However, most efforts ended up with diploid meiotic products.

3. Other students indicated that pairing of chromosomes took place without replication. In cases where the pairing was between homologues, this resulted in essentially a Zorkian meiosis with only one cell division. However, an interesting variation showed pairing between nonhomologues, and this led to strangely segregated daughter cells one of which was Aa and the other was Bb.

4. There were many cases in which students inverted the order of pairing and replication processes. This usually gave the correct meiotic outcomes, but based on an incorrect understanding. (Another illustration of how erroneous knowledge may not be a serious impediment to student success on traditional exam questions.)

5. Crossing over was commonly misrepresented. One mistake is based on an erroneous diagrammatic representation found in many first year and high school textbooks. In this model, two nonsister chromatids are shown physically "changing partners" and pairing up with their nonsisters. This erroneous representation is reinforced by the actual phrase "crossing over," which suggests that the chromatids do physically

change partners. Breakage and reunion may or may not be shown to occur after this. However, most of the time the outcomes are not affected by a "novel view of the process." Once again this is another example of the way in which a non-canonical viewpoint can give rise to a correct answer.

There were many other erroneous versions, almost impossible to classify. Similar confusion abounds among students in other courses. (For a comprehensive summary of students' ideas about meiosis, see Kindfield, 1994 and Lai, 1996.) Perhaps the most creative response observed in our course, was the "single strand model" of meiosis provided by one student in the post-exam test. A cell was drawn containing a single line (representing a chromosome); the single line replicated into two, which then segregated into two daughter cells. Not meiosis, but wonderfully creative and elegant in its simplicity. Surely such a system exists somewhere in the universe?

In closing, the point of this chapter is *not* that students are both creative and successful at scraping through the system. Rather, many of us who teach have failed to notice what our students are telling us about genetics and genetics education through their responses. There is much that can be learned about how to teach genetics by attending more closely to the finer details of what students do and say in our courses, and documenting the many routes to misunderstanding. Looking closely is akin to opening Pandora's Box.

References

Kindfield, A.C.H. (1994). Understanding a basic biological process: Expert and novice models of meiosis. *Science Education*, 78(3), 255–283.

Lai, P. (1996). *University students' conceptual understanding and application of meiosis*. Unpublished doctoral dissertation, 215 pp.

9

Does Problem Solving Work?

An article on problem solving written by John Sweller (1991) raises a tantalizing point about learning problem solving that is relevant to all those who are involved in the teaching of genetics. Sweller presented his point as a "myth" of education in the following way.

Myth: Practice at solving many conventional problems is an efficient way of gaining problem-solving expertise.

This so-called myth certainly seems to be the basis of a great deal of genetics instruction. Genetics professors present concepts generally in a lucid and interesting manner and then problems are assigned that are designed to test students' grasp of the material. The professor's pitch is something like, "If you understand these principles then you should be able to do these problems and then pass the exam." Right? However, something is clearly not right. Many students do not fit into the formula below:

$$\text{learn principles} + \text{do problems} = \text{pass exam}$$

Some step in the sequence is blocked and assessment becomes distorted. Furthermore, a common complaint from students goes something like the following: "I understand all of your lectures, I enjoy reading the text, and I feel that I truly understand the principles, but I still can't do the problems."

Sweller offers some interesting insights that may account for this all too common phenomenon. First he points out that educational research shows that experts tend to classify problems according to solution modes or "schemas," whereas novices are more likely to use superficial features to classify problems. For example, apparently the main difference between expert and novice chess players is not their ability to think ahead and consider all possible moves, but rather that the experts remember and can mentally access a large number of realistic game configurations (i.e., schemas), whereas for novices each game is a novel experience. In genetics these schemas would not be principles of genetics, but rather a repertoire of remembered pathways through similar genetic data sets. These pathways would have been learned either from the experience of actually doing genetic research, from manipulating genetic data, or from solving genetics problems in the past. In other words, we might be patting ourselves on the back just a little too hard when we solve a problem "from first principles." We might in fact be doing what many of us tell students *not* to do, which is to remember and apply set patterns of solutions that lie deeply embedded in our memory.

The solution to any given genetics problem may seem recognizable and obvious to us because of a complex arrangement of multiple, overlapping, and subtle features that we have organized and dealt with so often that they now simply represent a schema that resides subconsciously in our minds. The journey of travelling to the answer of the problem may have occurred so many times before in different but related forms that it is now, in a sense, automated. However, novices (students of genetics), because of their more limited experience, might try to classify the same problem on the basis of superficial and non-significant features such as "This is an *Aspergillus* question." Having identified and focussed on this particular structural feature, the novice may then attempt to solve all *Aspergillus* problems on the basis of mitotic crossing-over, regardless of whether they deal with mitotic crossing-over, mutation, or nondisjunction! Such a response may seem less surprising and more reasonable *if* we consider that the novice's only experience with *Aspergillus* is the small number of problems he or she has solved previously, most likely problems in mitotic crossing-over.

If we accept Sweller's thesis, what can be done to help students acquire their own genetic schemas? Clearly, dropping the students in at the deep end of problem solving would not be productive. The irony is that simply asking students to solve a lot of genetics

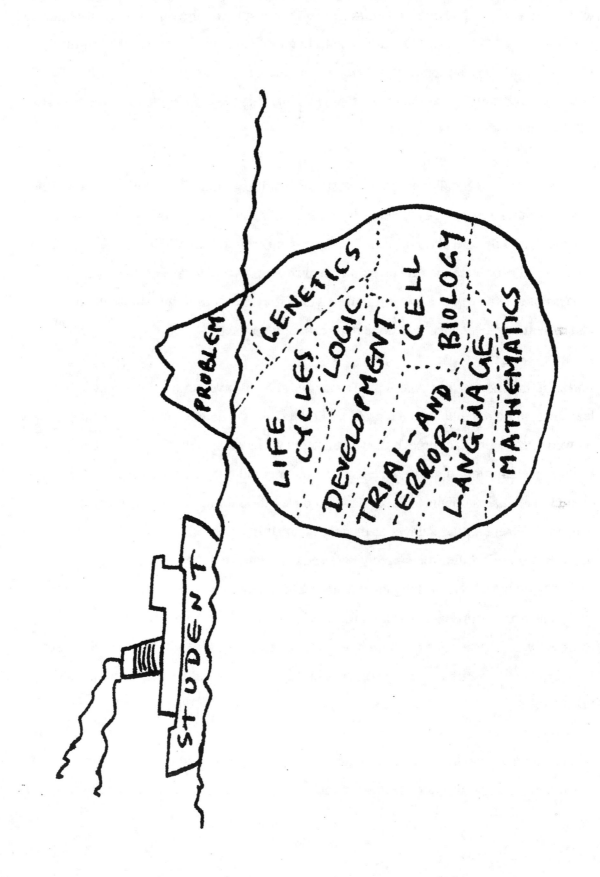

problems might not be the most effective way for them to learn problem solving! We
need to help novices focus their attention on recognizing and increasing their repertoire of
problem-states (i.e., develop schemas) rather than just focussing on superficial features
that might lead them down the garden path in their dedicated search for "the right
answer." Part of this process will involve convincing students not to "just solve
problems." Our experience suggests that changing this long-standing tradition requires
patience and hard work.

One idea from the Sweller article suggests that studying worked examples is a possible
way to acquire schemas. In genetics teaching this is a commonly used device, so little
more needs to be said about it, right? Wrong again…all too often when novices look at
such solved problems they focus *on* the answer without really carefully analyzing the
process of *getting to* the answer. In other words they aren't thinking about whether they
understand *why* each step was done in a particular manner.

Earlier in this book we argued that more active student involvement is needed in genetics
learning. So how can the solved problem be used as a vehicle for this? In particular, how
do we get students to approach the solved problem in ways other than just reading
through the solution to get to the answer? One method that has been used in other fields
of study is to ask students to explain a worked solution for a specific problem to
somebody else, another student or the instructor. It is now well documented that such
verbalization helps the student recognize and examine what they do and do not
understand about the item they are explaining. Further, the experience can contribute to
their repertoire of problem sets and, thus, help to establish thought patterns and,
eventually, schemas. (For ideas on how to organize such an activity in your class, see
Chapter Three on Think Aloud Problem Solving in Triads and Chapter Six on Models
and Mimics.)

A similar result can be obtained through "writing-on-the-reading" activities, in which the
student is asked to have a conversation (on paper) with the invisible author of the solved-

problem. The student's dialogue with the ghostly question-solver involves covering the printed prose with penciled questions, thoughts, and comments that reflect current ideas, misunderstandings, or pieces of inspiration. The process can be a methodical one where the student reflects upon and interrogates the author on each step in the problem. Alternatively a more holistic approach may be taken where the students would simply be exploring their understanding of the concepts presented in the problem. The results of this "writing-on-the-reading" activity can also be shared and lead to further discussion. (This teaching activity is an adaptation of the "Write on the Reading" procedure discussed in Baird & Northfield, 1992.)

Thus far we have considered ways of using worked examples as a means of helping students develop schemas by gaining more experience and expertise with genetic data sets. Another approach involves taking the traditional genetics problems we assign and using them in innovative ways to help our novices gain some of the "genetics experience" of the expert. Although we believe the possibilities here are limitless, we will consider two that we have experimented with. In each situation we discuss, note that the search for the right answer is secondary to understanding the nature of the problem and the approach to the problem. Getting students to accept the value of this may be the most challenging part of the task.

One idea to consider is *not* to make the correct solution of the problem the students' immediate goal; instead, the assigned problem is used to generate numerous tasks each designed to give the novice some practice with manipulating genetic concepts and data. The purpose of such solution-free tasks is to help the students focus on understanding the problem as a whole entity. This instructional device is what we can call "Problem Expansion." We think that the huge leap that students must take in moving from the genetics principles discussed in lecture and chapter to the solving of genetics problems can be eased with transitional questions based on the genetics problem itself. Every aspect of the problem can be expanded as ancillary questions that are posed to the student about the problem itself. To illustrate this procedure let us consider a sample genetics problem.

The Problem

John and Martha are contemplating having children but John's brother has galactosemia (an autosomal recessive disease), and Martha's great-grandmother also had galactosemia. Martha has a sister who has three children, none of whom have galactosemia. What is the probability that John and Martha's first child would have galactosemia?

Rather than rush into a solution, consider assigning a range of Problem Expansion questions as shown below. Note these are just meant to be representative examples of the many possible questions that can be asked.

1. Can the question be restated as a pedigree?
2. Can the question (or parts of it) be restated using Punnett squares?
3. Can the question (or parts of it) be restated using branch diagrams?
4. Can you make up a story about this family?
5. Define all the scientific terms in the question, plus any other terms that might give trouble to some (e.g., English as a Second Language) students—terms such as contemplating, great-grandfather.
6. What assumptions need to be made about this question?
7. Which unmentioned family members must be considered? Why?
8. What is the role of luck in the choices this couple has to make?
9. What are two generalities about autosomal recessive diseases in human populations?
10. In this family, whose genotypes are certain and whose are uncertain?
11. Sketch the relevant autosomes of each crucial individual in the family, and sketch meiosis diagrams showing the relevant genes.
12. In what way is John's side of the pedigree different from Martha's side? How does this affect your calculations?

13. Is there any given information that is irrelevant in the question as stated?
14. What can't you work out/determine from this problem?
15. How is solving this kind of a question similar/different from solving questions like question #N from the text (a problem unlike this one)?

The possibilities for this kind of expansion procedure are almost endless. Some questions are trivial, some are profound, but all help the student to develop schemas and to become metacognitive, i.e., to engage in reflection on the process of thinking about problem solving. An additional twist on problem expansion is to consider setting more general types of tasks before doing any specific questions. For example, consider asking

- What are the general things you need to consider when doing probability questions in genetics? Why?
- What are the problems one might encounter when doing probability questions in genetics?

After reading a specific question, ask

- Can the question be stated in other ways? (Get the students to suggest them.)

In addition, to emphasize the importance of focusing attention on the *process* of problem solving rather than the answer to problems, consider asking students to work through these types of tasks with a cluster of related problems *without solving them*, before moving on to other problems that they will solve.

A separate instructional approach involves using the genetics problems to focus the student on following a metacognitive pathway to the answer. In this case, understanding the *journey to the answer* becomes the goal. Ask your students to ask themselves a series of questions that require them to stop and reflect upon *what* it is they are doing and *why* they are doing it. For example they might use the following two questions throughout the process:

• What in this problem has indicated that I should take this next step?

• How will I know if this is the correct step to take?

In summary, Sweller's article reminds us that getting the answer to a problem is really not the main objective. Understanding genetics is the goal, and if we achieve this goal, then problem solving should be easy. Too often genetics instruction inadvertently puts the cart before the horse, and obsession with getting the right answers distracts from important learning processes that must occur for true mastery of the subject. The discussions above suggest that genetics problems can be used in many ways other than problem solving, and that these ways can assist students on their path to understanding genetics.

References

Baird, J.R. & Northfield, J.R. (Eds.) (1992). *Learning from the PEEL experience*. Melbourne, Australia: Monash University Printing Services.

Sweller, J. (1991). Some modern myths of cognition and instruction. In J.B. Biggs (Ed.), *Teaching for learning: The view from cognitive psychology*. Hawthorn, Victoria: Australian Council for Educational Research.

10

The Power of Concept Mapping

It is wonderful, if a calculation is made, how little the mind is actually used in the discharge of any profession.
Dr. Johnson, in Boswell's *Life of Johnson*

The mind, as Dr. Johnson noted, is a wonderful but lazy organ. The corridors of the mind are crowded daily with millions of bustling ideas, some significant and some not, but very few of these ideas enter the rooms that constitute the mind's library. There is an effective filter that allows only those ideas that seem necessary to pass. The key word is "seem" because, in fact, many necessary and useful ideas do not pass the filter because they do not seem necessary.

The contents of these "rooms" are laid down during a construction period in which a learner builds an understanding of a new concept. Once constructed, the door to the room is difficult to open. If new ideas appear that are in variance with or inconsistent with the constructs in the rooms, they are often ignored because the original construct seems to be inaccessible. We see obvious examples of this learning issue in genetics education. Often a conceptual model of meiosis constructed by a student during a biology class in grade school will persist unscathed up until university graduation and even into graduate school. If new ideas presented at university are not consistent with the student's original conceptual model, the model that is tucked away in its mental closet will not easily change.

However, clearly personal conceptual models do change, but equally clearly the process of change is ineffective for many students. If we accept the principle that all students are capable of meaningful learning, but some lack the necessary skills to do so, then we must ask what instructional processes can help in their task of deconstruction and reconstruction. This article is about a powerful and flexible teaching and learning tool that was designed to help students deconstruct and construct relevant mental models. This tool is called variously "concept mapping" (Novak & Gowin, 1984) or "mind mapping." It is a method that has been used extensively in grade school, but it is finding increasing use in undergraduate teaching in sciences and the arts. It is particularly appropriate for undergraduate teaching and learning in genetics because students find the content material of genetics so difficult to organize. Furthermore, students have great difficulty in deconstructing inadequate mental models of genetic principles. Concept mapping is a highly flexible and versatile educational device. Novak has successfully introduced concept mapping to large corporations who use it for product design and development. We believe that it is also a potentially useful device for professional experimental design. However, our goal here is to illustrate ways in which concept mapping can be used to further genetics understanding.

Concept mapping takes on several different forms depending on the intent of the exercise, but all these forms are based on the process of linking related concepts using clearly defined propositional statements. The result is a drawing with concepts spread on a page and joined by labeled directional lines. Whereas the process is straightforward, the outcomes are powerful. That being said, it is our experience that many people who are introduced to the basic idea of concept mapping categorize it prematurely as "too simple," "obvious," or "kindergarten-like." Therefore, when introducing concept mapping to a group of instructors or learners, a period of consciousness-level raising may be necessary before the process is embraced. However, often the challenge of drawing one's first concept map immediately dispels the notion of simplicity and illustrates the potential.

We will distinguish three different types of concept maps, which we believe have their own special uses.

1. Linking concept maps

In genetics, students have trouble relating the different levels of genetic analysis to each other (e.g., molecular, cellular, organismal, population). Also difficult is relating the different parts of the subject (e.g., gene interaction, Mendelian ratios, biochemical pathways, development). Linking maps are good for charting these kinds of relationships. Students are presented with a set of concepts (six to ten is a convenient number) and asked to interconnect them with lines that are labeled with a phrase that precisely defines the relationship between the terms being connected. (For a sample see the diagram on the opposite page.) To make the task more challenging and prompt a richer engagement with the concepts it is helpful if the terms used are a mixture of nouns (objects), verbs (actions), and processes (events). Try making a map of the following concepts

gene / allele / locus / segregation / heterozygote / 3:1 ratio

You will probably find that even though you think you know what the relationships are, putting precise words to them is a difficult or at least thought-provoking task. Some students react to this by claiming that they really do understand the relationship even though they cannot put it to words. We claim that the act of forcing students to put words to the relationships is a way of dragging the concepts out of their mind closets, dusting them off, and reassembling them as new or clarified constructs. Note also that even for six concepts as in the present example, there are a lot of interrelationships to be labeled (15 to be precise). It is important to try to label as many relationships as possible, even those that seem ridiculously remote, because this also promotes the metacognitive process (self-monitoring of ones own understanding).

An instructor can discover a great deal about student understanding by checking over students' mapping efforts. However, assessing and assigning marks to concepts maps of this type is challenging because there are an infinite number of "correct" relational

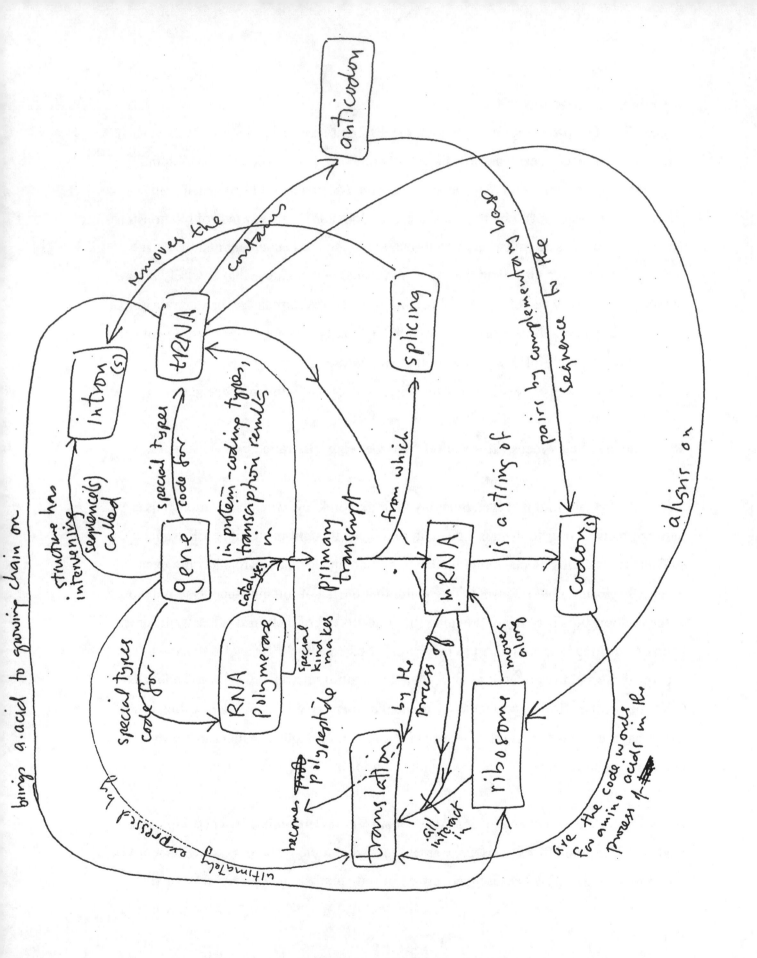

designs. Furthermore, because this educational procedure is a thinking tool meant to promote deconstruction and reconstruction on the part of the learner, any errors made are part of that process and do not deserve scrutiny for marks because this could deter risk-taking in describing the relationships. If concept maps are drawn in class or in tutorials, then a debriefing session in which different maps are compared is most illuminating. If your students are very familiar with concept mapping and have completed numerous maps, for discussion and "formative" purposes you may wish to collect and assess the level of understanding illustrated by linking statements. (See White & Gunstone, 1992, for ideas on how to do this.)

2. Hierarchical concept maps

Many students do not see a subject such as genetics (or even a subtopic such as gene interaction) from a personally integrated perspective. This was brought home to us after a recent genetics exam in which the students were allowed to prepare a single page "crib sheet" as a memory aid to take into the examination with them. The surprising feature of these crib sheets was that they were all pretty much identical! Furthermore, they were merely shrunken down versions of the lectures that the students had been given. In other words, the students had not processed or organized the information to the extent that they felt comfortable enough to prepare their personal view of the subject. They saw the subject merely as a linear stream of material that had flowed past them in lectures.

The role of the hierarchical map is to help learners reflect on and organize the subject material in a personally meaningful way. This time the mapping procedure begins by providing students with a single concept such as "mutation." In the first stage the students are then challenged either individually or in a group discussion to think up concepts that are related to the starting concept. In the next stage the students individually arrange the set of concepts into a hierarchy. The first concept is at the top with branches linking it to closely related concepts; from these secondary branches are links to the next level of the hierarchy, and so on. As with all concept maps, the lines must be clearly labeled with the connection idea written in such a way so that another person can understand it. An

71

alternative approach (which can be used if the one just described stalls) is to present the students with 10–20 different concepts and ask them to assemble these into a hierarchical map with labeled links.

Note that these maps are not meant to be flow charts. Although parts of the maps can be essentially sequential steps or processes, the intent is to chart the knowledge domains of the subject. Branches can be interconnected, bifurcated, or connect back to several levels, giving an extremely complex pattern that in no way resembles flow of material or information.

Hierarchical maps are extremely versatile educational tools and can be easily adapted for use in a variety of educational settings. Walk into any classroom from grade school up to the graduate level, and with one initial phrase or word you have a constructive and useful lesson activity. Moreover, it is this type of concept map that is used in business. We have heard of boardrooms whose walls are festooned with Post-it notes arranged as hierarchical branch diagrams showing all the different facets of marketing, production, testing, etc. of new products. Stories tell of the realization by companies that for an executive entering a new business or a new department, the "learning curves" needed to master the new job have shrunk from years to days through the use of such hierarchical concept maps.

3. Exploratory concept maps

These types of concept maps are used to try to chart the unknown. They provide a way for graduate students or professionals to explore a research area to determine what is known and what needs to be known. Students just starting out their careers as researchers in genetics or any other area of science often have trouble coming up with a research topic. The result is that in many cases the supervisor designates the thesis topic. The exploratory map is potentially useful to help students do this for themselves.

Exploratory maps can be generated individually or in a lab group meeting. One way would be to start with drawing a hierarchical map of the research area of the lab, for example beginning with the term "mitochondrial plasmids." The individual or group would then generate, draw, and link related topics. Thus, stemming from this first concept could come branches to inheritance, structure, function, distribution, and evolution. From these, secondary and tertiary levels could be charted. One of the benefits of this approach is that at the end of one branch a question mark will appear that will define an area needed for research. Perhaps even going beyond this, an association of concepts will become apparent that might lead to a Big Idea.

At the very least, practice with exploratory maps could save a great deal of embarrassment on comprehensive exams where graduate students suddenly discover large gaps in their command of their specific research area. Admittedly, some people can achieve this command without using a concept map, but many others can benefit from such a way of organizing their understanding.

We have learned from our experience using concept maps in many educational settings that they have great potential. However, to our knowledge, this educational tool is not being used much in university genetics teaching. In a difficult analytical discipline such as genetics, the area of learning in which students need the most help is not in acquiring the "facts" of a subject, but in using them. Concept mapping can help students deal with the challenge of bringing together their knowledge of different aspects of the discipline for use in applied settings and examinations.

To help you get started using concept maps we will close this chapter with some advice and a reminder. There are two key points we wish to emphasize about using concept mapping as an instructional device. First, concept mapping is *not* meant to serve as an information delivery technique. Maps are personal structures; thus an expert's map (i.e., one produced by the course professor) will often hold little value for a novice (i.e., the student). The underlying purpose and value in this educational procedure is the generation of the map by the learner. Our second point is that concept maps make explicit what is already known about the subject matter. Thus, learners must either have some prior knowledge of the subject matter being mapped or have assimilated some information about the topic before they can draw a concept map.

Finally, we reiterate the warning that many learners will balk at being asked to draw concept maps. For students who are comfortable using a passive type of learning the discomfort level of concept mapping may be a major deterrent to their acceptance of it as a valid educational process. Instructors will almost certainly sense this resistance and might not persist. Students who might have encountered concept maps early in grade school might reject them as "childish" devices. However, anyone who has actually tried to make a concept map in a complex subject like genetics would never call them childish. It is worth persevering because students have a lot to gain from concept mapping.

References

Novak, J., & Gowin, D.B. (1984). *Learning how to learn*. Cambridge: Cambridge University Press

White, R., & Gunstone, R. (1992). *Probing understanding*. London: Falmer Press.

11

Untethered Genetics

Any course of instruction (including genetics) is a self-contained microcosm, a universe unto itself. The course has its own content, its own principles, and its own standards for success. Furthermore the course hands out its own credential. A student "passes" the course by following an established set of procedures, and for this receives a permanent and lasting credential in genetics. At any time in the future, anyone who is curious could ask the student "Have you had a course in genetics?" If the answer is yes, the student is accorded all the intended rights associated with this, including prerequisites for other courses, qualifications for specific jobs, and of course entry into graduate school.

In short, instruction at colleges and universities is generally modular. The God of Universities at some stage divided all knowledge into "subjects" deemed to be self-contained, and then decreed that if a sufficient number of these is taken and passed, then the student is judged to have been educated, and is ready to assume the role of a useful member of society.

However, the modular approach is prone to several types of abuse. Perhaps most important is that success in the modules is often reduced to a type of game playing that is disturbingly detached from reality. Courses become untethered balloons of activities that have little connection to the real world, either the real world of academia or the real world of jobs or personal growth. Such balloons have floated through our ivy-covered halls for decades, and as long as everyone plays the game there is never the risk of an embarrassing pinprick. Students are adept at winkling out and learning the rules for success. They quickly find out "how to pass" or "how to get an A" in any particular course. In fact, several species of instructor evaluations ask the student specifically whether it was explained clearly to them what they had to do to pass the course, thereby encouraging this "how to pass" mentality. This contributes to a disastrous switch in

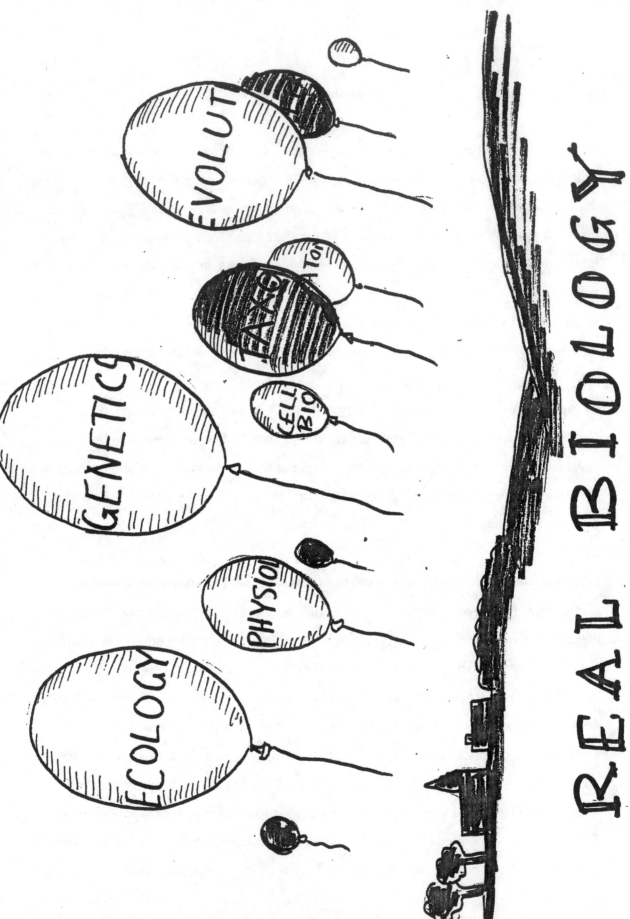

emphasis by the student. Rather than focussing on attaining a personal understanding of genetics, the student's goal becomes understanding the set of things that the professor expects of them in order to grant the valued credential. Thus, in genetics, as in most other courses, we have seen the rise of "credentialing" at the expense of learning.

Instructors like the modular approach to education, and play right along in the game. Here are some possible reasons for this behavior.

1. Course professors are busy trying to be both educator and researcher, so whatever streamlining and packaging is necessary to help in this regard inevitably gets done.
2. Instructors like the autonomy of being captain of their own ship. ("In MY course we do so-and-so"; "In MY course I always tell them this-and-that.")
3. Delivery and assessment of a given course is so much easier if there is a wall around the course and no "contaminating" material from other "subjects" is allowed in.
4. The game-playing model seems "so right" for everyone concerned. There is the "coach" (the instructor) and the "team" (the students), whom join together to try to maximize the number of touchdowns (A's) so that the Dean can congratulate all involved.

However, let's try a reality check. What are we really accomplishing if we all play this game? To illustrate this point, let's play a little game of our own. Let's look at a few examples of untethered knowledge, starting with the obviously ridiculous, moving into science, and then eventually our own backyard of genetics. The first example is a portion of a "course" module that provides a description of the mysterious substance traxoline. Read the passage below and then try to answer the questions that follow.

The montillation of traxoline

It is very important that you learn about traxoline. Traxoline is a new form of zionter. It is montilled in Ceristanna. The Ceristannians gristerlate large amounts of fevon and then bracter it to quasel traxoline. Traxoline may well be one of our most lukized snezlaus in the future because of our zionter lescelidge.

Directions: Answer the following questions in complete sentences. Be sure to use your best handwriting.

1. What is traxoline?
2. Where is traxoline montilled?
3. How is traxoline quaselled?
4. Why is it important to know about traxoline?

Admittedly, the above fabrication is okay for a bit of a giggle, but now for comparison, read the following module, which is based on part of a popular physics text.

Period of a plane pendulum with finite amplitude

In the limit of small oscillations a plane pendulum behaves like a harmonic oscillator and is isochronous, i.e., the frequency is independent of the amplitude. As the amplitude increases, however, the correct potential energy deviates from the harmonic oscillator form and the frequency shows a small dependence on the amplitude. The small difference between the potential energy and the harmonic oscillator limit can be considered as the perturbation Hamiltonian, and the shift in frequency derived from the time variation of the perturbed phase angle.

Directions: Answer the following questions in complete sentences. Be sure to use your best handwriting.

1. Under what conditions does a plane pendulum behave like a harmonic oscillator?
2. How does the oscillation change as the amplitude of a plane pendulum increases?
3. From what is the shift in frequency derived?
4. Which of the following can be considered as the perturbation Hamiltonian?

a) The large difference between the potential energy and the frequency limit

b) The small difference between the potential energy and the frequency limit

c) The large difference between the potential energy and the harmonic oscillator limit

d) The small difference between the potential energy and the harmonic oscillator limit

After reading these passages we decided that it was no wonder we became biologists! The second passage in particular may explain why the exit surveys completed by graduates from Ivy League universities show that most physics graduates could not explain how a pendulum works. They could all quote the formula for simple harmonic motion, but could not explain the mechanism in a way that reflected deep understanding.

This finding got us thinking about the curriculum of genetics. Surely this could not happen in genetics? However, consider the following material from a well-known genetics textbook. Then think about the questions that are posed below.

Unordered ascus analysis

Linkage may be deduced from the proportions of parental ditype, nonparental ditype, and tetratype unordered asci. For linked genes, at short distances values of parental ditypes are high (maximum 100%), tetratypes intermediate, and nonparental ditypes low (minimum 0%), but at the limit for large distances tetratypes are highest (maximum of 66.7%) and both parental ditypes and nonparental ditypes are equally low (minimum of 16.7% each). Thus for linked loci the value of the ratio nonparental ditypes / tetratypes lies between 0 and 1/4. In the case of unlinked genes parental ditypes are always equal to nonparental ditypes. The proportion of tetratypes is determined by the combined distances of the loci from their respective centromeres. For short distances parental ditypes and nonparental ditypes are relatively high (maximum 50% each) and tetratypes low (minimum 0%), and at the limit for large distances tetratypes are high (maximum 66.7%), and parental ditypes and nonparental ditypes are low (minimum of 16.7% each).

79

Thus for unlinked loci, the ratio of nonparental ditypes / tetratypes lies between 1/4 and infinity.

Questions to answer:

1. What is the maximum value ever reached by tetratypes?
2. What is the minimum value ever reached by parental ditypes?
3. For what linkage conditions are parental ditypes and nonparental ditypes equally high and tetratypes low?
4. What can you conclude if tetratypes are high and both parental ditypes and nonparental ditypes are equally low?
5. What type of linkage arrangement would give the following values: parental ditypes of 60%, tetratypes of 35% and nonparental ditypes of 5%?
6. What can you conclude if the ratio of nonparental ditypes to tetratypes is 16.7%?

All of these examples are challenging and require logical analysis, but are also totally without context. Thus, many modularized disciplines of university science suffer from verbal overkill and a low signal to noise ratio. The significance of these issues becomes apparent as we consider some of the implications of using this type of instructional material.

The first implication of building our instruction around Traxoline-like inscriptions is that we may be inadvertently training our students to be the type of robotic and rote learners we all love to hate. Students fed on a diet of Traxoline may assume what has been described by Paul Ramsden as a "surface approach" to learning (Ramsden, 1992), one symptom of which is student devotion to reading a text for "the answers" rather than reading for understanding. Problems associated with this surface approach to learning are well documented and discussed in some detail by Ramsden. The main pedagogical principle to note is that the nature of the tasks we set for our students really can influence how students learn.

A second and related implication of setting Traxoline-like tasks is that we may be leading our students down the garden path toward poor achievement, or worse, even failure. Students can successfully complete Traxoline-related tasks, but through the process will have acquired minimal understanding of the content or underlying principles. This can have disastrous effects in the long run, as deluded students will often study using these same techniques, then arrive blissfully unprepared for course examinations.

Our three examples of "disciplinary" modules illustrate one further issue—namely that language matters. It probably isn't apparent as we read through the genetics example, but the other two examples should make the point loud and clear that we speak a foreign language when we teach, and that most students (not only those for which English is a second language) must exert a great deal of energy learning the language as well as the principles. Thus as we lecture, model problems, or run discussions, our students may be lagging behind, translating the words we use, while simultaneously trying to make sense of the ideas we present. Most instructors never fully appreciate that the foreign language of their discipline must be practiced and mastered along with the content. This, however, takes time, a commodity in short supply in university courses. (Language issues that relate to the teaching of genetics are also discussed in Chapters Five and Seven.)

What can be done to prevent students from taking refuge as "surface learners"? Here are a few ideas to get you started.

1. Context is everything. Instruction always needs to have a solid footing in reality. Every analysis we present must be related to a basic biological question. In the case of tetrad analysis (for example) the sight lines from tetrads to all other related biology must be clear. Students must be able to draw a concept map that connects and interrelates tetrad analysis with more important biological issues, especially fungal life cycles, meiosis, recombination, variation, mapping, cell cycle, replication, DNA, mechanisms of crossing over, and evolution. Understanding and being able to explain these interactions are much more important than the tetrad analysis itself. The instructor needs to actively draw in

material from other courses to show these connections and students need to be provided with opportunities to practice making these connections for themselves.

2. The syllabus and our curriculum should no longer be modular. The term "genetics" today is only a convenience for pigeonholing certain pieces of biology. The pigeons must be let out of their holes! The borders between the traditional disciplines have become blurred. For example, genetics as a discipline does not really exist in biological research today. Instead, genetics is merely one of a number of techniques and analytical tools biologists draw upon to acquire understanding of the world around us.

What is needed is an integrated curriculum. We tend to see courses as solutions ("we need a new course in so-and-so") but we must learn that throwing new courses at a problem is not a solution. Maybe it is time to toss out the entire notion of having courses. Electronic technology makes it easier to conceive of a borderless curriculum founded on the big questions of biology. A less radical notion would be to introduce learning vehicles that are designed specifically to promote integration. Case-based and problem-based learning are ideal vehicles for examining broad questions that can only be answered by integrating knowledge from multiple sources. (For more information about these strategies see Chapter Thirteen.)

3. If courses are to be retained but designed in a more integrated manner, assessment must change. There is significant research to illustrate that assessment drives both instructional and learning approaches. Examinations and other forms of assessment must be congruent with our instructional practices. Thus, if we wish students to understand and respect the integrated nature of scientific knowledge, we must set tasks that assess that form of understanding. This must happen throughout our students' academic programs. Too often it isn't until third year that we set challenging tests that assess students' ability to think. By this point in their programs, students have settled into patterns of study and learning that are difficult to break. If we begin in year one of university education, then by fourth year a final assessment system could be set that

would be a rigorous test of integration skills. But, clearly for this to work the goals of the degree need significant redefinition.

4. Students need more experiences in real biology. We all acknowledge the virtues of field trips (yes, even for molecular biologists); non-cookbook, inquiry-based laboratories; and directed studies in research labs. Yet all of these activities are suffering from pressure of student numbers and dwindling budgets, and are being cut back. It is our experience that many biology students today are so unfamiliar with plants and animals that they cannot appreciate the great questions of biology that have fascinated the minds of others, let alone begin to develop their own questions. Despite budget restraints there are still some things that can be done.

Start simple. All biology begins with nature, so show the students some nature—many of them simply have no experience of it. Invite students from your class for a stroll across campus or to a nearby green area; talk about the trees, plants, animals, ecological succession, the original vegetation on campus before development, the impact of human activities on the biota; try some bird-watching; make lists of the top ten flowers, trees, mosses, birds, etc. Arrange visits to research facilities, and ask the denizens to explain their experimental set-up. Invite guests (faculty, graduate students and undergraduates doing research) into class to talk about what inspires them in biology. Ideally context should begin its construction in early childhood. One way to do this is through community-based science, i.e., encouraging children to observe and record living systems in their own backyards or communities.

The above ideas suggest that fairly radical changes are needed in our programs, planning, instruction, and assessment. Such large-scale change is intimidating, and, as we have learned from personal experience, often quite unproductive unless others around you are also changing. In order to move in a direction that ultimately leads to significant change, you will need to gain acceptance and develop credibility. We close this chapter with three suggestions that may aid you in making some forward progress. First, establish a small team of individuals who have some common vision, and then agree on the nature of the problem that needs to be addressed. Second, start small, develop a game plan, try out some new ideas; when you find strategies that work, develop a model that can be scaled up. Third, share your successes: it pays to advertise.

Reference and Endnotes

Ramsden, P. (1992). *Learning to teach in higher education*. Routledge: London.

A reference to the Traxoline module (attributed originally to Judith Lanier) can be found in an article by Charles W. Anderson (AAAS forum *Seizing Opportunities: Collaborating for Excellence in Teacher Preparation*, 1997).

The pendulum module was devised by Robert Cohen, Department of Physics, East Stroudsburg University, PA.

12

Learning Outcomes in Genetics

One of the biggest challenges for students is to absorb the blizzard of information delivered in a genetics course and distill it into a set of skills that will allow them to pass the course. The connection between the course content and assessment is perfectly clear to the instructor, but often not to the learner. Course material is organized into and presented as discrete subunits of instruction, but as each subunit is presented there is frequently little attention paid to specifics of the assessment process. From the learner's perspective it is not clear which parts of the course content represent essential core genetic knowledge and skills, and which parts are secondary. For example, all instructors like to tell stories describing the historical development of genetic ideas, and these are intended to enrich the course and facilitate understanding. However, although the students generally enjoy a story (and, as we have pointed out previously, story does provide important context that helps learners recognize the connections between concepts), they often cannot identify the key elements that will come back to haunt them on the exam. Students often complain that they have attended every lecture and read and understood every chapter in the textbook but they are still thrown by the content and style of the questions in the assignments and the exams. We appear to expect too much of students in their ability to separate the chaff from the wheat (and accordingly the noise from the signal). In our role as instructors we need to spend more time in making explicit the key elements that students must attend to when they prepare for assessment. By key elements we mean the essential learning outcomes of the course.

If students are presented with a list of learning outcomes they can be in no doubt about the instructor's expectations. Furthermore, the list can be used as a guide to help the instructor design assessment questions that more clearly reflect the priorities of the course. A list of learning outcomes may also serve diagnostic and remediation purposes later on as areas of weakness in student performance can be related back to specific objectives and show the instructor and the students where further work is needed.

In the 1970s and 1980s it was fashionable for science textbooks to include "learning goals" for each chapter. The idea was a pedagogically sound one because it focussed specifically on the elements discussed above. However, no genetics textbooks to our knowledge have ever adopted this practice. One possible reason for the absence of learning goals in texts is that it is quite difficult to pinpoint the learning outcomes in science generally and in genetics specifically. Precise and thoughtful wording is needed in order for the objectives to be useful in directing student learning. This precision could become cumbersome if taken too far, and some compromises have to be made to avoid statements that smack of "legalese." The point is that it is much easier for a busy instructor not to get involved in this kind of writing, and simply let the students figure out the learning objectives for themselves.

Despite the obvious pedagogical potential of well-defined learning outcomes, some instructors balk at the notion of introducing these to their course. Such educators are uncomfortable with the defining of learning outcomes because they point out that if used with "fundamentalist fervor" the list of objectives can lead to the uninspired practices of teaching specific behaviors or only the verbatim learning outcomes and no more. We wish to make it clear here that we are *not* advocates or supporters of such a "back-to-basics" notion of genetics teaching.

In what follows we have provided one possible list of learning outcomes for core material in genetics. The list is structured on the topic sequence that is used in one of our single-term courses in basic genetics at the University of British Columbia. The list is not meant to be complete, but rather to provide a set of examples of what genetics learning objectives might look like. The lists are arranged in groups under headings that represent major topics taught in our course. In writing the outcomes we have tried to stay away from the admonition "Understand so-and-so" because this is unspecific and students undoubtedly would not know the criteria that are needed to demonstrate understanding. To make this list a bit more universal there are places where our wording is somewhat unspecific. If you wish to adopt these we suggest that you modify the language to suit your specific course needs. All the items on the list are written in a manner that indicates a type of action or performance by the student that might be used during assessment. Some items are highly quantitative and require analysis (problem solving), whereas others are descriptive and require some type of explanation.

List of learning outcomes

1. Review of biochemistry and cell biology

  • Draw a cell diagram showing the nucleus, chromosomes, rough and smooth ER, ribosomes, Golgi, secretory vesicles, lysosomes, mitochondria, chloroplast, membranes.

  • Draw an interphase chromosome complement with major landmarks: centromeres, telomeres, nucleolar organizer.

  • Draw a typical eukaryotic gene with major landmarks: coding region (exons), introns, regulatory regions, RNA, and protein start and stop sites.

  • Draw a short stretch of a DNA molecule labeling 5' and 3' ends.

  • Draw a short stretch of a protein, labeling amino and carboxyl ends.

  • Draw a diagram of transcription showing template and nontemplate strands with 5' and 3' ends of DNA and RNA.

  • Draw a diagram of translation, labeling 5' and 3' ends of RNA, and COOH and NH$_2$ ends of protein.

  • Draw a cell diagram showing the flow of genetic information from gene to protein including proteins used in cytosol, mitochondria, chloroplast, and secreted proteins.

  • Draw a series of cell diagrams illustrating chromosome movement at mitosis. List main chromosomal and cellular events of PMAT.

• Draw a series of cell diagrams illustrating chromosome movement at meiosis. List events of PMAT1, PMAT2.

• Diagram the eukaryotic cell cycle.

• Diagram DNA replication showing leading strand, lagging strand, Okazaki fragments, 5', 3', ligase, DNA polymerase.

• Label stages at which mitosis and meiosis occur in diagrams of life cycles of 2n, n, and 2n/n organisms.

• Label stages at which DNA replication occurs in diagrams of life cycles of 2n, n, and 2n/n organisms.

• Draw a diagram showing how DNA replication takes place during chromatid formation and label old and newly synthesized strands.

2. Mendelian genetics

• Diagram the experimental sequence whereby it can be shown that discrete alternative phenotypes are determined by the alleles of a single gene.

• Use the principle of equal segregation to predict progeny of crosses of known genotypes for a single gene.

• Use the principles of independent assortment and equal segregation, and the product and sum rules of statistics to predict progeny ratios from independent genes.

• From progeny ratios, deduce parental genotypes.

• Demonstrate the use of branch diagrams and Punnett squares in predicting progeny genotypes and phenotypes.

• Use pedigrees to deduce autosomal dominant and recessive inheritance of human disorders.

• List three examples each of autosomal dominant and recessive disorders in humans.

• In a pedigree, use the laws of inheritance for an autosomal gene and the product and sum rules to calculate risk of affected children in a specific mating.

CORE GENETICS

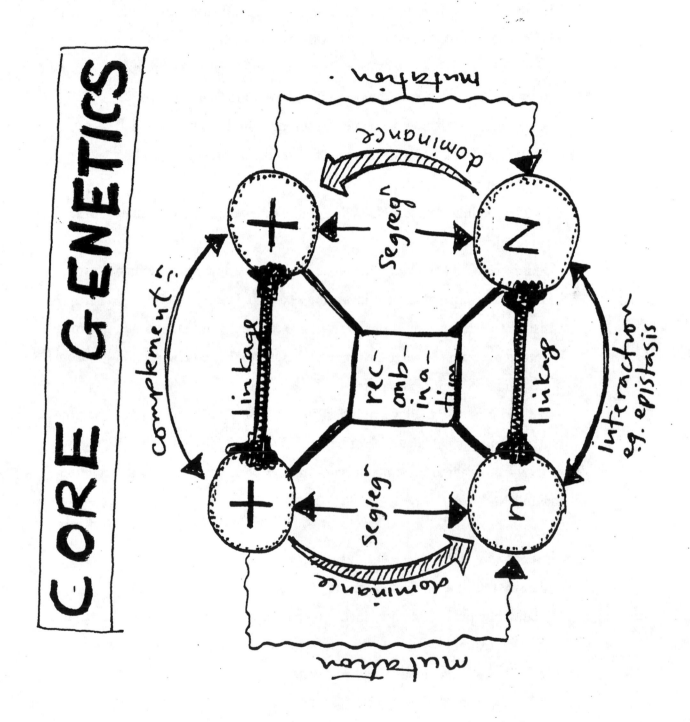

3. Chromosome theory of heredity

  • List three examples each of unique genes, gene families, and repetitive DNA.

  • Diagram the structure of nuclear organizer, telomere, centromere.

  • Draw a diagram that illustrates the problem of replicating telomeres.

  • Draw a nucleosome, and higher levels of chromatin condensation.

  • Draw diagrams showing the chromosomal conformation and the segregation of two unlinked heterozygous autosomal genes at mitosis and meiosis.

  • Draw diagrams that demonstrate the chromosomal basis of Mendel's laws.

  • List two differences in the inheritance patterns of autosomes and sex chromosomes.

  • Deduce the inheritance of X-linked dominant and recessive human disorders from pedigrees.

  • Deduce sex linkage from inheritance data in experimental organisms.

  • Predict outcomes of crosses involving sex-linked genes.

  • List three examples of X-linked recessive and one dominant human disorder.

  • Draw a diagram showing the organization of sex-linked and pseudo-autosomal genes on the X and Y chromosomes.

  • In a pedigree, use the laws of inheritance rules for an X-linked gene and product and sum rules to calculate risk of having affected children in a specific mating.

  • Draw a diagram to illustrate mosaicism through X chromosome inactivation in mammals.

4. Gene interaction

  • Identify multiple allelism from progeny ratios in a set of crosses.

  • Predict outcomes of crosses involving parents heterozygous for multiple alleles.

  • List two examples showing the interaction of genotype with environment.

  • Describe two functional tests for allelism.

  • Infer dominance/recessiveness from cross data.

  • From given data, distinguish full, incomplete and co-dominance.

• Infer the type of gene interaction implied by 9:7, 9:4:3, 12:3:1, 13:3, 15:1 ratios in F2.

• Predict phenotypic ratios from given modes of gene interaction.

• Draw a concept map relating a modified F2 ratio to biochemical pathway interactions.

• Deduce incomplete penetrance and expressivity from breeding data.

5. Linkage

• Deduce linkage/none in a diploid testcross.

• Deduce linkage/none in a haploid cross.

• Predict results of testcrosses of dihybrids (or haploids differing at two loci) in cases of specified linkage and of no linkage.

• Deduce gene order, map distances, and interference in a diploid 3-point testcross.

• Deduce linkage/no linkage from data from a selfed trans-dihybrid.

• Draw a diagram that illustrates understanding of the nature of RFLPs/SSLPs and their detection.

• Diagram the use of RFLPs/SSLPs in linkage analysis.

• Deduce linkage of a human disease locus to an RFLP/SSLP locus.

• Deduce chromosomal locations of genes from human/rodent hybrid cell data.

6. Mutation

• Invent symbols for dominant and recessive mutations and their wild-type alleles.

• Diagram a detection method for mutations in a haploid and in a diploid.

• Diagram a selection method for mutations in a haploid and in a diploid.

• Draw a diagram that distinguishes somatic from germinal mutation.

• Diagram the consequences of germinal and somatic mutation.

• Calculate mutation rate from fluctuation test data.

• Distinguish reversion from suppression using cross data.

• Draw a concept map relating the terms proto-oncogene, oncogene, somatic mutation, cancer.

• Diagram the process of mutational dissection.

• State the diagnostics of a mutation caused by transposon insertion.

• Diagram the procedure of transposon tagging.

• List three categories of DNA lesions that cause gene mutations.

• Draw a concept map interrelating the concepts of mutational lesion, gene action, mRNA, protein, mutant phenotype.

7. Chromosome rearrangements

• Distinguish gene mutation from chromosomal mutation using given data.

• From data deduce the presence of the main types of chromosomal rearrangements including inversions, translocations, duplications, and deletions.

• Diagram meiosis in heterozygotes for an inversion, translocation, deletion, and duplication.

• List the main effects of heterozygous chromosome rearrangements on linkage.

• Map genes to inversion and translocation breakpoints.

• Use deletions for chromosome mapping.

8. Changes in chromosome number

• List three examples of human aneuploids, their karyotypes and phenotypes.

• Diagram meiotic nondisjunction (first and second division).

• Using genetic markers, deduce first or second division nondisjunction.

• List the diagnostics for detecting aneuploidy in a haploid and a diploid experimental organism.

• Predict the outcome of a cross of a trisomic heterozygote at one locus.

• List three genetic conditions that give rise to chromosomal imbalance in progeny, and diagram each process.

• Infer the existence of polyploidy from data.

• Diagram mechanisms of auto- and allopolyploidy.

• Predict outcomes of crosses of allopolyploids to parents.

• Calculate segregation ratios for heterozygous loci in autopolyploids.

• Diagram process of making synthetic allopolyploids in plant breeding via a sexual method and cell culture method.

9. Gene-protein-phenotype relationships

• Describe an experiment that shows that DNA is the genetic material.

• Describe an experiment that shows that genes code for proteins.

• Distinguish complementation from recombination in data that show wild types arising from the union of two mutants.

• Draw a concept map that relates the terms wild-type allele, wild-type phenotype, missense mutation, nonsense mutation, frame-shift mutation, amino-acid substitution, protein truncation, null function, leaky, mutant phenotype, genetic disease.

• From data, distinguish loss-of-function mutations from gain-of-function mutations and give an example of each.

• From data, deduce order of mutant sites in a gene by flanking marker combinations of intragenic recombinants.

• Use prototroph data to map mutant sites in a gene.

• Use DNA sequence data to map mutant sites in a gene.

• Draw a concept map that relates the terms epistatic gene, hypostatic gene, sequential hierarchy, developmental pathway, phenotype.

• Draw diagrams that illustrate the consequences of mutations in regulatory region, coding sequence, introns.

• Describe two experiment tests that distinguish between a structural mutation and a regulatory mutation.

10. Population genetics

• Calculate allele frequencies from genotype frequencies.

• Determine if a population is in HWE for a specific autosomal locus.

93

- Calculate allele frequencies from phenotype frequencies, assuming HWE.
- Calculate genotype frequencies from allele frequencies assuming HWE.
- Calculate mating frequencies from allele or genotype frequencies.
- Calculate allele frequencies after one or many generations of given intensities of selection against recessive or dominant phenotypes.
- Calculate mutational equilibrium from forward and reverse mutation rates.
- Calculate mutation/selection equilibrium values of allele frequency form given values of w and u, and vice versa.
- Determine equilibrium genotype frequencies under various levels of inbreeding.
- Determine if two loci are in linkage equilibrium.

11. Quantitative genetics

- Calculate various statistics relevant to quantitative variation, e.g., variance, correlation, regression.
- Calculate broad sense heritability from F1 and F2 variance, and from correlation between relatives.
- Calculate narrow sense heritability from selection differential.
- Discuss the relevance of heritability measurements to quantitative human behavioral phenotypes.
- Identify QTLs from data.
- Map a QTL from data.
- Calculate phenotypic ratios of phenotypes determined by polygenes/QTLs.

After assembling and re-reading this list we realized that although this is only part of our genetics curriculum (we do the remainder in the subsequent term) it was a surprisingly long list, especially when one considers that each item on the list requires careful application of quite intensive analysis. The problem is heightened by the fact that for most students just about everything on the list is new. Although they might have heard of the concepts before, they have not been expected to demonstrate "flexible performance capacity" around these topics using deep understanding and considerable analytical poise (see Chapter Four for further discussion on flexible performance capacity). This realization highlighted for us the significance and value of delineating and emphasizing learning outcomes. Genetics is perhaps the biggest challenge that students will encounter in their biology degrees. We must be sure to present the subject as a coordinated set of incisive principles, and not a jumble of bewildering analyses. The list provided here also clearly points to the danger of potential curriculum overload—even a modest course content such as the one shown is full of immense challenges for the beginning student.

We must stress again that the outcomes provided here are intended to serve as a working list and not meant to be chiseled in stone. An outcomes list should flex and evolve with your course. Consider using your list to aid beginning teaching assistants in designing tutorial experiences. Any item could become a jumping off point for whole class or small group discussion. And, most importantly when you devise and customize your own list, be sure to share it with your students. Lists of learning outcomes are meant to be public documents that aid all participants in the educational enterprise.

13

Introducing Problem-Based Learning to Genetics Education

This chapter considers the application of a problem-based learning approach to the teaching of genetics. We propose below that Problem-Based Learning (PBL), sometimes known as case-based learning, is potentially a powerful instructional approach for achieving many of the goals of science education in general, and of genetics education in particular.

Before we begin our discussion, we need to clarify a few points for those unfamiliar with the notion of Problem-Based Learning. We'll begin with an illustration of what PBL is *not*. Many genetics instructors try to put across the principles of genetics using problems, but this does not mean that they are using Problem-Based Learning. In the lecture theater, problems are used as illustrations of principles, or modeled as examples of how a geneticist operates "in practice." Unfortunately, this lecture arrangement is well intentioned but wrong-minded as it is the lecturer, and genetics expert, who is doing the problem solving. Traditional lecture format deprives students of something they desperately need, which is the opportunity for acquiring relevant scientific data themselves, and practice manipulating it and coming up with meaningful interpretations. "Good" lectures do all this work for the student, and the students remain passive and inadequately engaged with the material.

What then is Problem-Based Learning? In its original form PBL is an instructional method that uses real world cases or problems as vehicles for students to learn critical thinking, problem solving, and the fundamental ideas of the course. Note that the emphasis here is on learning and not teaching. PBL is conducted in small groups of less than 20 students. Students actively engage with the cases and build their own understanding under the guidance of the instructor, but the instructor does not (and cannot) do the building for them. Thus, PBL is an instructional approach that in its quintessential form completely does away with lectures as sessions of information transfer. A PBL approach is perfectly in keeping with the constructivist view that deep understanding and learning come from active processing of new information; new ideas are constructed when the learner compares new information with previous mental constructs, revealing concordances and discordances. Identification and acknowledgment of discordances by the student is a particularly important part of the process of learning because these serve as "sticking points" and areas of confusion that must be clarified before new ideas can replace old.

The first widespread application of PBL was in medical teaching faculties (McMaster University's medical school was one of the pioneers in this). It is easy to see how instructors in medicine can present a practical medical case for the students to grapple with, and that this will have immediate relevance to the students who want to become physicians. Commerce departments also use PBL; once again a manifestly practical problem is presented such as how to price a new beverage that some firm has developed for sale.

In PBL settings the students discuss the case as a group and bring to bear all their existing knowledge about it. Then they begin to define "learning issues," which are more or less what they do not know and need to know in order to solve the case. The learning issues are prioritized, and then a decision is made about which students will deal with which issues, and if there are any principles or concepts that the whole group needs to deal with. Learning resources are considered, and here the instructor and students decide on where they can find the relevant information.

At the next session, each student teaches the rest of the class what they have learned about their assigned issue. Attempts are made to integrate the new information, and to relate it to previous knowledge. If the learning process raises new questions, these too are listed and the cycle is repeated until a satisfactory evaluation of the case can be made. New information or data (but not answers) may be provided at this point by the instructor to further guide the students' investigations. In all of these processes the instructor attempts not to inform but to guide, support, and encourage the students' initiatives. Instructional faculty typically need to adopt a radically different view of the educational process and special training sessions are provided to prepare instructors for the new roles and responsibilities they assume in PBL classrooms.

The basic PBL format described above was designed with highly practical real world problems in mind, of the type particularly prevalent in professional schools. Can genetics educators in faculties of science use the same methods? Indeed we are hearing about success stories of using PBL in undergraduate instruction, mainly in the basic sciences. But the format needs some modification for each specific science instructional setting.

The success of PBL rests upon choosing and/or designing the right type of case study. In applied genetics it would be relatively straightforward to devise "real need" problems very much like those used for PBL in medical schools. For example, one could set the challenge as trying to increase the amount of glucoamylase (an enzyme used commercially to convert starch to sugar) produced by the commercially important fungus *Aspergillus niger*, or to increase the amount of the drug penicillin produced by *Penicillium* species. For a plant-based problem, the goal could be to devise a strategy for producing the anticancer compound taxol using yew tree cell culture methods (to avoid harvesting endangered yew trees). Even the "curiosity-driven" areas of genetics (and science generally) lend themselves well to PBL cases. For example, how could one design a research program to develop a detailed genomic linkage map of a 'new' experimental organism such as some specific marine alga or a rotifer? In addition there seem to be many opportunities for useful problem designs presented by trying to clone various genes of interest to basic research. Designing experiments to investigate various genetic phenomena is also an obvious avenue to explore. Newspaper articles on genetic topics (such as those found in *The New York Times* or the *Globe and Mail* science pages) could generate interesting problems. All of these challenges require extensive literature research by students to check on organismal biology, life cycles, and appropriate genetic approaches. In fact, the point is that science is just as practical as medicine and commerce, and the act of carrying out the PBL exercise effectively models the necessary steps that all scientists have to go through to carry out their experimental programs.

Can the types of genetics problems found in genetics textbooks be used in PBL? These problems are in themselves microcosms of research, and generally represent actual data derived from a genetic experiment. Therefore they could indeed be used with a PBL format. However, remember that the PBL approach is not the same as just assigning problems for students to do as homework. The strategy requires that the same steps be followed as for the 'real world' problems. In other words the problems should be newly presented, discussed in small groups, and lists of learning objectives devised by the students. Ideally the challenge should be open-ended, reasonably complex and multidimensional, and with no obvious answer. The individual research period might be shorter and of a different type; probably many of the learning issues could be dealt with using the textbook right there in the classroom. Finally the students would teach each other what they discovered, and discuss the integration of their newfound ideas in order to solve the problem. In short, adapting PBL in the genetics classrooms is another method for giving students the opportunity to use an active cooperative approach (such as they do in many research settings), and move away from the passive listening-and-copying approach.

PBL methods work well in small classes, but many instructors find themselves in the thick of mass education, and are required to process hundreds of students in their course. Obviously one way of dealing with this problem in order to try PBL is to divide up the class into smaller groups, but in most cases this simply is not possible; the instructor does not have enough hours in the day, or there is no money to hire TAs, and so on. So is it possible to apply the PBL approach in a single classroom containing hundreds of students? We suspect that the basic form of PBL cannot be used effectively in this setting. There simply is not an adequate opportunity for effective discussion and interaction; furthermore the instructor cannot monitor and support PBL in such an unwieldy group. However, the spirit of PBL can be alive and well in such classes.

Although the logistics of discussion seem impossible, if thoughtfully structured, student interaction in the large group lecture setting is both feasible and productive (we speak from experience). It is possible to use the large group classroom setting for active data processing along the lines of PBL. One way is to present a learning module (say a genetics principle) and then present a problem (or another type of limited data set) that is related to that principle. Then go through some of the PBL steps. First ask students in a short quiet period to prepare a list of items that are needed to analyze the data (provide some exemplars to get the process started). Call for some of these from the class but don't let this go on too long or some students will get restless. Then allow another short quiet time for students to analyze the data individually. Afterward have a debriefing period in which you ask students for key pieces of insight that allowed them to advance through the problem. If none are forthcoming, present your way of solving the problem. Then ask for questions on your solution. If the silence is deafening at first, don't despair. Change in student participation happens gradually with encouragement. Finally another problem of similar structure should be presented, and this should present no difficulty. In this way the class proceeds as a stop/start series of instructor monologues interspersed with periods of active student processing.

Assessment is an interesting challenge in PBL. Ideally the instructor completes the assessment during the sessions and no exam is ever held. If academic policy at your institution requires an examination then at the very least your assessment instrument should be congruent with the PBL approach. Assuming that including a PBL library research session is not feasible as part of an exam format, then other elements could be included. For example, students could be given a problem and asked to provide the list of items needed to solve the problem and a commentary on the analytical steps used or attempted. Some instructors who use PBL approaches include a question at the end of the exam that is designed to be answered by research groups working together. If these types of questions are not possible or palatable, then the PBL sessions can be viewed merely as a means of getting the students mentally adept at the subject, and therefore able to pass any exam. However, if you want your students to respect and value the time spent in PBL, your assessment strategy should reflect your instructional approach.

Assessment is just one of the challenges. Another hurdle is development of effective case problems. This requires thinking about your course content differently and experimenting with problem design. Expect that some revision will be needed and this may take time and reflection on your part. The good news is that PBL is taking off in a number of disciplines, and dedicated PBLers are now amassing compendia of problems that are accessible via the Internet.

Reports of using PBL in the medical education community suggest that the potential gains do justify the effort. PBL-trained individuals are a match for their traditionally taught peers in standard exam settings. Furthermore, PBL graduates appear to have gained a deeper and richer understanding of concepts and retain their knowledge longer. Success in the real world that looms after the University degree seldom demands graduates who are walking encyclopedias of science. Quite the opposite is true. Society needs individuals who are creative thinkers and problem solvers, people who can communicate their ideas well, and who are good interactive team players. PBL is well equipped to prepare students for to meet these demands.

14

Genetics Education and Life-Long Learning

It is our impression that most undergraduate science students have only a hazy vision of how their present studies relate to their future lives. There is some feeling that graduating from a degree program will lead to a better career, but the precise connection is vague. Furthermore, there is an impression that the real action only begins after they receive their degree and leave the university (the "seat of learning"). These students regard facts as hard currency, synonymous with "knowledge," whereas the skills associated with writing, discourse, and social interaction are often downgraded and viewed as a waste of time, rather than attributes that will be useful throughout their lives.

We saw a recent illustration of this during a particularly difficult session in a genetics lecture. The scenario began when the students were asked to participate by discussing with their neighbors a set of genetic data displayed on the overhead projector. Previously in the course the students had been exposed to an array of teaching and learning "devices" designed to encourage, motivate, and entice even the most reticent of students to speak. So they knew the instructor placed high value on these activities. Nevertheless, except for a few hushed whispers, on this particular day almost nobody was talking, or discussing their insights with their neighbors, or volunteering ideas—the students just sat there and seemed to be waiting for the "real" lecture to begin. In frustration, the lecturer mentioned that he knew most of the students were intent on medical school and therefore they might be interested in a conversation he had had the previous night at an awards dinner, seated next to a member of the medical school admissions committee (all true). This person had reported the very interesting fact that the prime reason why qualified applicants are rejected from medical school is because of poor communication skills. Instantly the level of participation in the genetics class increased dramatically. Students turned to their neighbors and engaged actively in debate of the data, people started volunteering answers and points of view, and there was a great sense of energy in the room. Sadly, the parable was soon forgotten; at the next lecture period the class was back to its usual sluggish self and participation in the data analysis was practically nonexistent. Apparently the habits and predispositions of a lifetime of instructor-centered education could not be overcome by one finger-wagging experience. Preparing themselves for life-long learning was not uppermost in the minds of most students.

Nevertheless life-long learning is a notion that seems to have become very popular among educators recently—it even has acquired its own acronym (LLL). Why the sudden surge of interest? Hasn't LLL always been important? As is often the case, the answer is Yes and No. Although the idea of life-long learning always draws a favorable response, the vision that most people have when they hear the phrase is of a hobby style of learning; learning a new language, for example, in order to broaden one's horizons ("another language, another life"). Nobody questions the merit of continuing this type of educational activity throughout life, and indeed one could argue that such activity is of paramount importance for promoting personal enjoyment of the arts, literature, and music, etc.—in short "getting a life." This, however, is only one aspect of LLL. When educators speak of LLL they are using the term more broadly. Implicit in this broader view is not only the humanitarian notion that a life-long learner will have a fulfilling life, but also the instrumental notion that LLL will help individuals maintain professional currency and thus contribute to their careers. This aspect of LLL has not always been considered important in education. We suspect that the recent renewal of attention directed toward LLL stems from a harsher side of reality, which reflects a more urgent need to keep pace with the alarmingly rapid rate of societal change that is occurring, both in the workplace and in personal lives. Let's look at some of these changes and see how undergraduate teaching and undergraduate learning methods, particularly in genetics, measure up to the new demands being put upon them by society.

Foremost among recent social trends is the shift to an increasingly information-based society. Anyone who has looked for specific genetic items on the World Wide Web knows that there are thousands of sites out there. Whereas some are trivial, a large number contain detailed and up-to-date information about many aspects of genetics. However, working with the WWW is very different from working in a library. On the Web, students have immediate access to a vast range of authoritative and up-to-date material that would be almost inaccessible through a library. It seems obvious, then, that the Internet and other information technology cannot and should not be ignored; these resources will constitute a significant part of the knowledge base that people will need to use and interact with on a continuing basis throughout their lives. We argue, then, that an understanding of how to access this information technology should form an integral part of our teaching curriculum. However, an equally important factor is the evaluation of the material found on web sites. Many of the sites sound appealingly authoritative, but what criteria can be used to assess this? We all know of cases of documents, electronic and not, which despite the inclusion of extensive bibliographies from peer-reviewed journals nevertheless espouse views verging on the crackpot. This suggests that another supporting skill we should emphasize in our curricula for LLL is how to critically evaluate information obtained via new media.

To what extent can the Internet be used as a educational tool in courses? To explore ways to help students gain experience with this technology we assigned a genetics Web-surfing project to students enrolled in the University of British Columbia's genetics course, for a total of 5% of the course marks. From this we learned some interesting facts. First, we found out that about half the students were completely unfamiliar with the technology and had to be taught how to use it from scratch. Second, we discovered that our own university was hopelessly equipped for providing Internet access to students. The first problem is shrinking as our undergraduates now arrive with a much more extensive background in technology. Nevertheless, it seems likely that institutions generally are still at a primitive stage in coming to terms with the new medium. Keeping up with the information explosion even within a relatively narrow field of interest has become a big challenge for institutions, professionals, and students alike. Most of us are gaining skills, but still are not very good at it. Learning about methods of dealing with this technology must become a required part of curriculum if we hope to inspire and support LLL.

Along with the information explosion have come explosions in other types of technologies. The point is not that students need to be trained in the most recent technologies (this is being done, although patchily), but rather that preparedness for repetitive retraining has to become a regular part of the career path—an ongoing part of the life-long process. No longer will people operate a lathe, or a stamping machine, or even a DNA sequencer for their whole lives. Their skills will need to evolve, and willingness and facility to learn new skills must be a life-long commitment. Large- and small-scale economic forces are already reforming the workplace in an unprecedented manner, and there is every expectation of a revolution in work styles over the coming decades. In addition, new occupations and careers are emerging that require not only new technologies but also new attitudes and skills. For example, internationalization will likely become a force in most occupations, requiring very different types of approaches to work ethics, technology transfer, and interpersonal communications.

What curriculum can we provide that will equip our students for flexibility and adaptability in their professional (and personal) lives? Is genetics education useful in this regard? Call us optimists, but despite the rather discouraging anecdote we shared previously we believe that a case can be made that learning the intellectual discipline of genetics and the problem-solving approaches of geneticists (rather than rote learning of scientific facts) will equip our students for this type of long-term advancement through life. In their future lives our students will need to be flexible problem solvers that can communicate and cooperate with others in the workplace and society. The key is that students need to acquire and recognize the value of these fundamental learning attributes, attitudes, and skills that will be portable throughout life and applicable in a range of situations. Students need to regard these as their core practices, essential for LLL, rather than as a set of superficial behaviors that will be sloughed off after graduation like a snake's skin.

What are the attributes, skills, and attitudes that encourage LLL? What role can genetics education play in promoting LLL?

First, as we have noted above, an individual must be literate concerning the critical use of information and global information technology. Genetics, by its very currency and rapid pace of change provides rich opportunities for promoting this type of information processing. The rapidly increasing collection of DNA sequences is a good example. Learning how to access, compare, and analyze these data requires nimble use of information technology.

Second, people must be armed with an arsenal of learning strategies. We have described and assessed a number of such strategies and their relevance to learning genetics in previous portions of this book, notably those sections that focus on problem solving, concept mapping, multiple representations, problem expansion, and zooming between different organizational levels (molecule to population). Each of these approaches emphasizes the fundamental role of discourse in learning, the importance of being able to reflect upon and evaluate information, and one's own understanding of that information. As a result of such approaches students can learn how to distinguish deep from shallow understanding, and can make creative connections in life. In our experience one of the rarest talents in undergraduates and graduate students is the ability to forge connections between different parts of their educational curriculum (what we can call "helicopter vision"), and of course this should be an essential component of LLL. Perhaps what sets genetics apart is that by its nature it needs to be explored widely and deeply to make any sense of it; success in genetics requires helicopter vision.

Third, individuals should have well-honed interpersonal skills. This includes a sense of personal agency, autonomy, and the ability to relate to and work with others. Problem-based learning, think aloud pair problem solving, Triads (see Chapters Three and Thirteen), and other group learning experiences in the context of genetic data analysis all provide excellent exposure to these skills. These are the skills that life-long learners will need to draw upon on a daily basis if they are operate in and move between the fast-paced career tracks of today's technology-information-bombarded society.

The fourth and last skill we mention is an inquiring mind. The best way of inspiring students to take an inquiring attitude toward life is through personal example. We all remember well the teachers who inspired us by showing us the things that motivate them.

- We remember the field trip in the pouring rain led by the botanist who, although clearly crazy, nevertheless showed us entirely new ways of looking at nature.
- We remember the teacher who actually showed us the dynamic world under the microscope lens. (Regrettably, both labs and field courses are dying breeds, the first items to disappear in the wake of budget cutbacks and downsizing.)
- We remember the lecturer who showed us how to delve into the structure of the invisible gene through mutation, complementation, high resolution mapping, and sequencing.
- We remember the old curmudgeonly professor who insisted that we must have good reasons for believing what we believe, and that all else must be rejected.

There is no substitute for such passionate states of mind in inspiring long-lasting commitments to learning.

Science provides not only a way of finding out about the universe or the cell, but also of finding out about oneself, humanity, and society. To engage in scientific discourse requires many levels of communication skills that form the fundamental basis of LLL. Regrettably, such experiences that require self-directed, experiential, and reflective learning are rare in the University environment. But designs for improvement exist—and one model for this is right under our scientific noses. In a book by Bereiter and Scardamelia (1993) on the nature and implications of expertise, the authors suggest that what takes place within productive scientific research departments or faculties may serve as a new model for educational systems that develop communities of learners and thinkers. They refer to this new vision of education as a "knowledge-building community." According to Bereiter and Scardamilia, a learning community is characterized by the following:

1. The sustained study of topics in depth rather than broad and superficial coverage.
2. A focus on problems rather than categories of knowledge.
3. Inquiry driven by students' questions.
4. Student explanation and critical assessment of ideas and theories.
5. An emphasis on discourse to generate joint knowledge and understanding among participants.

Their approach to learning aims to produce expert-like generic learners that leave their university experiences with a vision and a purpose to strive for expertise and learning throughout their careers. Obviously, both radical rethinking of what takes place in university classrooms, and far-reaching curriculum changes are needed to encourage the practices suggested by Bereiter and Scardamelia. These changes must become part of a general plan, rather than the brave efforts of inspired instructors in a few courses. The very nature of courses and instruction itself will need revision to make the university the lasting educational experience that students will remember, value, and draw upon as they prepare to become life-long learners.

We suggest that the thoughtful design of genetics courses can provide a useful framework for developing many of the types of skills and attributes necessary for life-long learning. At the core of this statement is a belief in the power of problem solving through data analysis. To engage in analysis takes intellectual poise, courage, and confidence. To communicate one's involvement and understanding of the analysis process requires carefully honed skills in discourse and a critical evaluation of information derived from a variety of sources. The student must be able to say with confidence, "I believe this because this data tells me it must be so."

If we wish to cultivate life-long learners in our genetics (and other undergraduate science) classrooms, instructors need to come to grips with the invisible problems in our teaching practices that fly in the face of what is needed to support and develop life-long learning. A few of these problems are presented below in the following abbreviated "List of Lamentable Lapses" (LLL) taken from a larger list provided by Phillip Candy during a recent visit to our university.

- Overloading the curriculum.
- Imposing too much detail at too advanced a level.
- Making excessive use of lectures and other didactic approaches.
- Failing to connect learning to the world of practice.
- Using forms of assessment that encourage "reproductive" learning.

A quotation by Pace (1971) seems a fitting way to end this chapter:

> A university is a habitat, a society, a community, an environment, an ecosystem. It should be judged by the quality of life that it fosters, the opportunities for experience and exploration it provides, the concern for growth, for enrichment and for culture that it exemplifies. The question is not just "What does your machine produce?" but also "How does your garden grow?"

References and Endnote

Bereiter, C., & Scardamelia, M. (1993). *Surpassing ourselves: An inquiry into the nature and implication of expertise.* Chicago: Open Court.

Pace, C.R. (1971). *Thoughts on evaluation in higher education.* Iowa City: The American College Testing Program.

This chapter was inspired by a lecture given by Philip Candy at the University of British Columbia. For additional information on the concept of life-long learning see

Candy, P.C., Creert, G., & O'Leary, P. (1994). *Developing lifelong learners through undergraduate education.* Commissioned Report #28, National Board of Employment, Education and Training, Australian Govt. Publishing.

15

Less is More

Ideas about teaching emerge when you least expect them. In 1995, I (AG) sat writing out the answer key for the three-hour Christmas exam in our Biology Department's core genetics course. Teaching genetics for the semester was over and I was preparing the answer key at the invigilation table during the examination itself, while 500 students labored away doing the problems. Surprisingly, making up the answer guide took 80 minutes. Although this was a task that had been done routinely for many years, for some reason that particular reiteration of this traditional Christmas exam period activity led to an epiphany. The revelation was that even though the instructor is an "expert" in the topics being tested and *knows* the answers to the ten problems on the exam, writing these answers out in a clear and logical way took almost half of the time the students had been allotted to both come up with the answers and to write them out. It became clear then that completion of the exam in the allotted three hours is an unreasonable expectation for students who have spent only one semester learning new concepts and new skills to analyze complex data, and in a course that students find the most difficult in the life sciences. Real science never imposes such a deadline upon creativity. Yet in "school science," situations like this are almost the norm.

Since that cold morning in 1995 we have taken a closer look at our expectations of students in the course. Although the exam tradition could not be eliminated entirely, we decided to establish a more realistic objective for our students. We altered the number of questions that the students are required to answer in the exam periods. More specifically, we reduced the number of questions from ten to six on the three-hour final examination, and made proportional cuts (from three to two) in the two one-hour midterms. Furthermore, the course content has been pared down and more processing time provided in class, in tutorials, and in the Genetics Student Help Center. The subject matter dealt with in the course is at the same or a greater depth, but the coverage is narrower. That is, we eliminated selected topics from the curriculum.

After these changes were implemented we noticed that students no longer complain about feeling rushed or having inadequate time to complete examination questions. Performance has also improved, in that the course average has increased and failure rates have dropped. Regrettably, there has been no way to set up a control group for comparison, so it is impossible to determine if the better performance we have observed is due to the narrower and deeper coverage.

We use this anecdote as an entry into discussing the demands that are put upon today's post-secondary students, not only in genetics but also in the university in general. We believe these demands are excessive and that university science (and humanities) education would benefit from a reduction and refocusing of this pressure. The principle "less is more" is the central theme of this chapter. It is a concept that seems to be attracting the attention of post-secondary educators across the world.

We live in an age of the greatest-ever explosion of information in the history of human beings, and this seems to be the driving force behind the overstuffed curricula that we see in today's academy. As more knowledge is gained, then the course curricula expand accordingly. A glance at the increasing size of introductory biology textbooks over the past thirty years provides a physical representation of the problem. University and college science professors justify this gluttony for content in terms of "rigor." Rigor according to *The Oxford Concise Dictionary* refers to severity, strictness, and harsh measure. Is such rigor an appropriate goal for an institution of higher learning? We believe it is not, and should not be, the main objective that university science instructors are hoping to accomplish in their courses. Rather we seek and expect coherence and precision in argument, and as argued elsewhere in this book a capacity for flexible performance (see Chapter Four).

As a group, university professors pay homage to the need to educate students in the ability to think and to integrate ideas from different parts of their programs, but our actions speak louder than our words. In most science courses taught in universities, the reality is that students have little time at their disposal to process and reflect on the material that is being thrust at them. So whereas as scientists and professors we value thinking, reflection, and coherent communication and argument, we do not teach in ways that permit these skills to be developed and practiced. In the vast majority of university science courses, students neither have the luxury of time nor adequate pedagogical structures in place to encourage them to talk about their ideas to their colleagues, nor to devise and try out new ideas on their instructors.

The problem of overload becomes compounded at the programmatic level because *every* professor feels the same way about *his or her* course and curriculum. University science students typically enroll in five "rigorous" courses every semester. Their days are filled with hardwired schedules, and their evenings by the outrageous demands of completing detailed lab write-ups and other assignments. The documented result is that rather than learning, our students engage in "studenting" (Fenstemacher, 1994) and what a recent Doonesbury cartoon referred to as "credentialing." Little wonder that the products of this process, rather than being well-rounded scholars who will be thoughtful members of society, sometimes seem to be shallow goal-driven automatons. So, we reap what we sow. In courses where harsh measures are the standard for success, it seems fitting that the metaphor for the science degree program has become "Survival."

Why have we let this happen? Who has made these educative and programmatic decisions? Who has let curriculum overload become the norm? Let's consider genetics as the example. What constitutes a well-rounded graduate in genetics, someone who could go on in the field of genetics and not be an embarrassment to him- or herself and to their institution? In deciding the phenotype of the ideal graduate, we must distinguish between the student goals of learning genetics, and learning to do genetics. Whereas the implicit assumption for many years was that we were training people for a career *in* science, it is now clear that only a small minority will take up a career in any aspect of science. Even though some "how to do science" is implicit in any scientific education, as educators we often confuse the two goals and sometimes understanding pays the price of too much emphasis on the turns and twists of practical genetics.

What do we expect in a portable kit for life-long appreciation of genetics? We might all agree that our graduates should know how to analyze inheritance patterns and understand the DNA/RNA/protein trinity. How much more do we require? What about cloning and analyzing Southerns, Northerns, and Westerns? Perhaps not quite so important. What about RFLP and PCR analysis? Even less crucial? Yeast two-hybrid system? (Less?) DNA chips? (Even less again?) Somewhere the line must be drawn—but where? One could argue that the core principles of genetics were already established 30 years ago, and that all that has emerged since then is detail. Whether true or not, it is likely we can all agree that the rate of addition of new principles of genetics is slow. (Test yourself—What are three new principles of genetics that have emerged in the last 10 years? A difficult question to answer? OK, make it three principles of biology.) What has increased, indisputably, is the amount of genetic technology. But, how do we decide how much technology should be in the lifelong genetics kit beyond the relatively compact set of cloning, sequencing, PCR, molecular markers, and the various blotting techniques? Keeping abreast of technology is important because it is the key to understanding why we believe ideas to be true, but a smattering of technology goes a long way.

Another force that causes course content to swell is what we term here as the "content ratchet" effect. A ratchet is a mechanical device that allows change only in one direction. The content ratchet works in the following way. There is always an incentive to add material to the course coverage because this imparts a feeling of up-to-dateness and rigor. But we fall in love with our own knowledge and lecture notes. Their contents become old friends that we enjoy introducing to the students year after year. We even post our notes on the Web and point to them with pride. However, excising material is a less attractive proposition. It seems so unrigorous to delete or simplify.

A similar ratchet principle appears to be the force behind monster textbooks. There are two parts to this. First, textbooks are reviewed by multiple reviewers, many of whom point to the necessity to include this or that, or want more explanation of something else. The second part is a concern for maintaining the *appearance* of a rigorous text. Another anecdote will illustrate this point. At a nearby university the biology instructors met to select the text for the coming year. There was a concern raised that the "encyclopedic" text currently in use was inadequate. There was simply too much material for the students to read and digest in one year, and students couldn't see the forest for the trees. Consequently it seemed there was a good case for adopting a shorter, more conceptual text that focused more clearly on the principles. However, some instructors at the meeting argued that the encyclopedia would always be useful in second and third years as a reference work. Others argued that their students had used a large comprehensive text in their high school biology program, so it would look very bad if a university first year course adopted a shorter text than the one used in the secondary school. This was a tale that got straight to the heart of the matter: science professors have the illusion that more is better, that robust is effective.

Striving for currency is also potentially a problem. The inclusion of the latest stuff as course material might be counterfeit currency in the sense that it doesn't contribute to the students' cognitive bank accounts. It looks robust to outsiders looking in (and administrators), and it seems substantial (in terms of notebook pages), but it doesn't add value as understanding.

Albert Einstein's writing is a great source of pithy quotes, and the following three are relevant to the issues we have raised here:

> "The only thing that interferes with my learning is my education."
> "Imagination is more important than knowledge."
> "Education is what remains after one has forgotten everything learned in school."

In these quotes Einstein seems to be indirectly referring to that most precious commodity of science (and society): creativity. Of course it is difficult to be creative in the absence of knowledge. We are definitely not advocating that content knowledge be set aside or thrown away in favor of teaching courses in creative thinking. The question is how much knowledge is needed to act as the substrate for the creative process? In our opinion, again the answer is "less is more."

The overstuffing principle is active at the curriculum level. Most North American universities have hit upon a curriculum that is composed of a diet of about five courses per twelve week term over four years. It is not clear how this number has come about, but it is an instructive question to ask what would be the effect of halving this number? Would this produce a better or a worse qualified graduate? Why don't we think about setting some "cognitive load limits"? This couldn't help but improve a student's experience in strolling through the cool marble halls of learning. Aiming for deep conceptual understanding of a more selective curriculum might even improve the exit test results of science graduates in the United States, many of whom have difficulty explaining even the most elementary of scientific concepts on tests of basic conceptual physics (Mazur, 1999; Hestenes & Wells, 1992)

The same sort of question could be asked at the course level. What would be the effect of halving the number of chapters covered in our course texts, and examining each concept more thoughtfully? Students are given more opportunity to explore the material, to think about it, to relate it to other concepts learned, and ultimately to use the material creatively.

What constructive steps could be taken by instructors wishing to reform their course content? Just like student change, transforming the practice of the instructor does not come easily, or overnight, or as a result of reading a small book on teaching genetics. It is a stepping stone in a long journey that must be made from geneticist to genetics educator. First the need for change must be perceived, and internalized. Any change undertaken must be feasible in the context of the program and of the course. Sudden and dramatic changes in practice and programming will upset the balance of many factors, so taking things one step at a time is probably a good idea.

We offer some suggestions below for those wishing to rethink their genetics curriculum in line with a "less is more" philosophy. Some of the ideas are reasonably obvious, but have been included anyway because we regard them as fundamental to the change process.

1. Avoid letting the textbook determine the content and sequence of your course. Set your own story line.
2. Draw a concept map of your subject area—find five key and essential topics and build your curricular plan around these.
3. Prepare a list of learning goals for the course (see Chapter Twelve) and make sure that your instruction addresses these.
4. Select the textbook chapters you wish the students to read. Read these yourself, noting how much time is required. Then review the chapters. Eliminate all the frills (for now) and assign only the essential pages.
5. For your selected list of essential topics decide what you need to do to help the learner understand these topics. Then, design an array of metacognitive activities for your students that will help them learn (some specific ideas for this are present in our other chapters).
6. In preparing assignments look for a close match between assigned problems and the key topics in your concept map and learning goals. Do every problem you plan to assign—write out the full answers in full sentences, noting the time required. Try to eliminate problems that overlap and cover the same topics (save these problems for practice and discussion in class).

In closing, partly for fun, but also to start you thinking about your personal journey, you might find it interesting to draw a concept map (or discuss with a colleague or students) interrelating the following terms used in this chapter: rigor, currency, metacognition, studenting, credentialing, curriculum overload, and less is more.

Learning a complex subject like genetics is not easy and students should not find it an easy ride, but that does not imply that it should be an ordeal. Leisurely reflection, library work, and discussion are all important parts of scholarship and therefore in the education of scholars. These activities are not possible on the stress-filled academic treadmills that constitute many curricula and courses in biology today.

References

Fenstemacher, G. (April, 1994). *Studenting: Promoting an institutional focus on improving college teaching*. Paper presented at the annual meeting of the American Association for Educational Research, New Orleans, LA.

Hestenes, D., & Wells, M. (1992). A mechanics baseline test. *The Physics Teacher*, 159–166.

Mazur, E. (1999). *Problems of University Physics Education*. A presentation to Physics Department, University of British Columbia, Vancouver, Canada.

16

The Five Golden Ways

Courses of instruction are inevitably linear. That is, some topic is introduced on the first day, and new and related topics are continually introduced until the course is over. However, the structure of a scientific discipline such as genetics is not wholly linear, and neither is its practice. Although some elements of the subject are linear (e.g., certain types of experimental procedures such as mutational dissection), the professional takes a holistic view of the subject, expertly integrating the levels of molecules, chromosomes, cells, organisms, and populations. Thus, in its practice genetics is more of a complex web or net. In planning such a course, the instructor is always faced with a dilemma about the most appropriate sequence with which to introduce the topics. In other words, how the holistic and integrated view of genetics can best be illustrated in a linear teaching and learning sequence. Whereas there is no one correct answer, in this chapter we consider five alternative curricular sequences that could be used in courses on introductory genetics to provide an integrated look at the discipline of genetics. These five alternatives or "Five Golden Ways" are described below, classified for simplicity by the topic that comes first.

1. Mendel First

This "historical approach" is probably the one that is most used by lecturers who are themselves professional geneticists. Because genetics is a young discipline, it is relatively easy to chart the main experiments that led to advances in understanding the nature of genes, gene transmission, and gene action. Mendel is the "father" of genetics, so the method must begin with Mendel's experiments, which were performed in the middle of the nineteenth century. The historical approach is well suited to the linearity of course structure because history is basically linear; that is, the events of one year by definition precede the events of the next. The first stages of the history of genetics involved the discovery of genes, their inheritance patterns, their dominance relations, their association with chromosomes, their mapping, and their interaction. All these occurred roughly in a linear sequence that is suitable for linear course structures. However, because each seminal discovery by definition leads to a host of secondary discoveries, the strictly linear treatment rapidly becomes complicated by the numerous secondary and tertiary-branched pathways down which these discoveries lead. The subject moves ahead on many fronts like the branching growth of a fungus, and all these branches need to be carried forward pedagogically if the strict historical approach is to be maintained. It is at this stage that, for the sake of a nice story, instructors might be tempted to speculate on linear sequences that in reality might have not been valid historically. It is truly impossible to convey the complex intellectual interactions that must have taken place between researchers proceeding down parallel paths. Hence the linearity tends to break down somewhere in the middle of the course.

Nevertheless, there are several important benefits of teaching genetics using the historical approach.

a) An historical approach attempts to convey to students the way that science works. Real people design real experiments to investigate specific areas of ignorance. The approach shows that science is not infallible and that scientific understanding is a human construction rather than a concrete and unquestionable fact. The flair for experimental design and interpretation are priceless commodities in science. Whereas not all can be Nobel prize winners, the seminal experiments that move the discipline ahead deserve to be held up as role models for young people, many of whom wish to pursue careers in science.

b) The historical approach to genetics instruction places all state-of-the-art knowledge at the apex of a pyramid of historical experimentation underlying it. Hence it conveys well the sense that the discoveries of any scientist are made possible because they "stand on the shoulders of giants." How disappointing it is to talk to young scientists who do not know the historical underpinnings for their current beliefs. Surely one of the jobs of an education in science, whether for budding scientists or just as part of liberal education, is for students to develop the habit of automatically asking themselves the question 'Why do I believe this to be true?' Such self-questioning is a practice that we regard as an important lifelong learning attribute.

c) An historical approach inevitably builds from simple to difficult because the discipline of genetics naturally expanded in size and complexity through time. Hence the approach seems tailor-made for building student comprehension. Also, the historical sequence has been progressively reductionist, starting with observable phenotypic variation and ending with detailed molecular characterization. This direction mirrors the way that most people start thinking about the world, starting with general observations about reality then proceeding to ask for progressively more detailed explanations.

d) For students with considerable previous exposure to genetic ideas, the historical slant can provide a different way of looking at what might be considered by the student to be a rehash of previous experiences.

So what are the limitations of such a widespread and popular approach to genetics instruction? First, while starting with what was known, such as observable phenotypic variation and ending with detailed molecular characterization may mirror how our knowledge has unfolded historically, it may not be the easiest way to develop a deep understanding of our current knowledge base in genetics. Each classic experiment we explain provides an episodic memory for the learner that includes salient information about people, concepts, symbols, and interpretations. The novice will mentally file all of this information unselectively, lacking any knowledge about which details are significant and need to be retained as permanent principles and which details may be ignored. Following presentation of an historic episode, we add contemporary explanations of some "older" data and reinterpret the historical findings, and in doing this we often add more facts and modified symbols that require some previous details be dropped or downplayed. For the expert this may seem enriching and help provide the full picture, but for the novice these modifications and extensions may make it difficult to separate the "signal" from the "noise." By trying to enrich, we may unwittingly increase a student's confusion.

Second, an historic approach may not necessarily help all students understand how to perform genetic analysis. As instructors, time is our enemy. We must constantly decide what is most difficult for students to understand, and what kinds of activities will assist students in gaining a deep understanding of the discipline. Time invested in explaining various traditional viewpoints and historical experiments might best be reallocated toward helping students understand what is known today and using these principles as the framework for examining and analyzing data in the modern context. Whereas students can usually retell the story of Mendel, they have much more difficulty interpreting a set of data to deduce monohybrid or dihybrid segregation.

2. DNA First

Because all life is based on DNA, it makes sense that we should be able to conceptually build up all the levels of genetics from an initial description of DNA, its organization, replication, and function. There is no doubt that this method has great application to students trying to integrate all the facets of the subject and "see the forest for the trees." Rather than getting bogged down in historical details, a course designed around this curricular sequence gets right down to business from the word *go* and provides a sense of modernity and relevance (because DNA is a "hot topic," appearing regularly in newspaper and TV articles).

However, the initially appealing linearity of this approach gets stuck in some twists and turns. For example, when discussing DNA it would be advantageous to talk about modern ways of analyzing DNA. However, this really cannot be accomplished without first learning about bacterial genetics, which is needed as a prelude to a discussion of recombinant DNA technology. Thus the instructor following this course approach, like one adopting the historical approach, will also have to make some important decisions about which of a number of alternative "linear" paths to take.

The DNA First approach does, however, liberate instructors to adopt virtually any sequence they please after the initial introduction to DNA. It provides the luxury of speculating on the ideal pedagogical sequence. Of course, any genetics course must cover the "top ten list" of genetics topics, which are listed below.

Top Ten Topics

gene and genome structure

gene function

gene transmission

gene recombination

gene interaction

gene mutation

chromosome mutation

genes and development

quantitative traits

genes in populations

Inspection of this list shows that there is no ideal pedagogical order even if one takes a DNA First approach in their course design. To be pedagogically ideal each topic should build on the next in a sequence that starts simply and increases in complexity step-by-step. However, it is obvious that several combinations of the "top ten" go hand in hand— for example gene function, gene interaction, and gene mutation.

But perhaps the biggest danger of the DNA First approach is that it seems to be particularly prone to become didactic. In other words, this approach can easily degenerate into the "mug and jug" style of education we mentioned in Chapter One. Filling students with the wonders of our subject matter— "DNA replicates like this...," "Transcription occurs like this...," "Genes are inherited like this..."—while initially appealing, leads to shallow and transitory understanding for reasons that we have detailed before. We reiterate that, no matter how clear the delivery, students will not necessarily accommodate the new information into their private mental constructs unless they are given the opportunity actively to process the material (deconstruct and construct).

Finally, those instructors adopting a DNA First approach must also be on guard against the tendency to teach genetics as a steady stream of facts with no historical or intellectual footings. This would be the worst kind of dogmatic non-science, leaving out all the doubts and inconsistencies that form part of the essential methodology of science.

The method does however provide an opportunity to devise "modern" footings for the principles. For example, the instructor must decide on the *best* (not the historical) evidence that genes are parts of chromosomes, or that DNA is the genetic material, and that genes code for proteins, and so on.

3. Mutation First

Most genetic analysis performed by professional geneticists begins with some type of variation, induced ultimately by mutation. Hence the Mutation First approach is the one that best mimics the science of genetics as practiced. However the professional approach is a difficult one with which to inspire students. Geneticists are only in the game because they have some burning biological question that needs to be answered by the incisiveness of genetic analysis. Few students have been inspired with such questions at the undergraduate level, so the mutational approach must seem irrelevant to their present or future lives. It is the same as saying, "If you ever have aphids on your roses, the following method would be the best way to kill them." Without ownership of the problem, that is, having an aphid infected rose bush as a motivator, the solution (however effective) is hollow and lacks meaning.

The mutational approach does well on the linearity criterion because although a professional would analyze mutants by several methods simultaneously, there is no reason why these cannot be introduced one at a time—inheritance, allelism, complementation, mapping, cloning, population studies, DNA phylogenies, etc. One problem with this sequence is that the motivation for much of genetics is to understand some aspect of development. So developmental genetics could rightly be the first topic, and a simple treatment of development could be given at the outset, and revisited once the complexities of genetics eventually are unfurled. The drawback of this approach is similar to one mentioned for the Mendel First approach—revisiting topics can lead students to become confused when they try to determine if principles they were taught early on in a course are still relevant or simply "noise." Population geneticists might, however, challenge a Mutation First sequence that begins with the topic of development. They could argue that variation is a population phenomenon so populations should be the jumping off point for the course, and this also makes sense in terms of the way that genetics is practiced. This clearly underscores the point that no one linear sequence will satisfy every pedagogical requirement.

The practical Mutation First approach is well suited to problem based learning (see Chapter Thirteen on PBL).

4. Simple Organisms First

The notion here is that simple organisms are easier to understand than complex diploids like humans or fruit flies. They have simpler genomes (smaller and uncomplicated by diploidy and repetitive DNA) and go through simple life cycles not involving that immense conceptual iceberg that has sunk many a student Titanic—meiosis.

Concepts of gene structure and function, dominance, gene transmission and recombination, complementation and cloning can be demonstrated easily in prokaryotes and viruses. Then eukaryotes can be introduced using haploids such as fungi as examples. In fungal crosses there is only one meiosis to worry about, but meiosis itself represents a considerable step up in complexity. Diploids can be introduced last, at a time when most of the basic concepts have been introduced, and in principle it should be a simple matter to extend the previously established concepts to their more complex life cycles.

There is no doubt that the linear aspect of this approach is appealing. Pedagogically it makes sense to go from simple to complex, so in principle the method is sound. However the biggest stumbling block seems to be that students generally have great difficulties understanding unfamiliar life cycles. For example the analysis of tetrads and mitotic crossing over in fungi are topics that represent an immense hurdle in most courses. This is not because the genetic processes are difficult (they are nothing more than allelic segregation and recombination) but because the life cycles are unfamiliar to students, so the simple processes cannot easily be placed into their proper perspective. Indeed these cycles are often a challenge to professional geneticists not working with fungi. Most students enter a genetics course with a relatively firm understanding of diploid life cycles; they possess knowledge about having babies, plant seeds, eggs and sperm, and they at least know where meiosis takes place in the organism. This is an enormous advantage to introducing diploids first because the students can concentrate on the genetics without having to learn about reproductive cycles.

5. Human Genetics First

Human genetics is one of the biggest "hooks" we can use to motivate genetics students. Genetic diseases and polymorphisms are of great interest to most students, even those not bound for medical school. Therefore why not use this hook to full advantage and introduce human cases right away to model genetic processes? Large doses of this approach are certainly appropriate for non-science majors, and moderate doses are appropriate for science majors. Genetics can provide great insight into the nature of our humanity, and human genetics is directly relevant to many aspects of everyday life. Students read about new discoveries and ethical dilemmas in medical genetics daily in the newspapers, and most lives are touched directly by some type of genetic disease.

However the limitation of this approach is that students will end up with a distorted view of scientific progress. Many science writers describing advances in medical genetics convey the erroneous idea that the intellectual driving forces of genetics are centered only in the fields of medicine and economics. Recent fussing over sheep cloning and

126

genetically engineered crops are good examples, raising headlines like "Should genetic research be banned?" (Whereas this headline refers to certain types of controversial genetic research, the phrasing comes as a shock to most geneticists.) Whereas we concede that much excellent information has come from research in these areas of social concern, the fact is that most of the significant conceptual advances in genetics have come not from studies on humans, sheep, or soybeans, but from research conducted on the model organisms like *Drosophila, E. coli,* and *Neurospora*. Applications of such advances to medicine, industry, and agriculture are the fruits of a healthy tree of basic genetic research. This is an important idea for students to carry with them into their future careers, genetics-oriented or not.

So although each of the Five Golden Ways has its own gilded features, none is perfect. The relative success of each pedagogical plan inevitably depends on the precise setting of a course within the curriculum of study. Depending on the institution, genetics is taught at the first, second, or third years of post-secondary education. Over this three-year period there are extensive differences among learners in terms of attitudes toward learning, previous exposure to the subject matter, and level of intellectual maturity. Most genetics courses are aimed at science majors, but a significant proportion is aimed at students in the arts or other faculties, and these different audiences demand different approaches. Previous experience in genetics is also heavily dependent on high school exposure and information gleaned from television, newspapers, and magazines. All of these factors add up to widely differing skill levels and motivations. By the time students reach the university classroom there are relatively few "blank slates" waiting to be inscribed by our instruction. Most students already know a great deal about genetics, but the depth and breadth of their understanding is highly variable. At one extreme of the range, we find students who have got the facts muddled or wrong, and will need to expunge these notions in order for deep learning to take place. In the middle of the range we find incomplete and shallow understanding. At the opposite end of the range we find possibly the most challenging attitude, that of students who have been exposed to much of the course content in their earlier educational experiences and who think they understand it.

This is possibly the most difficult attitude to deal with because with this attitude comes closed-mindedness and an unwillingness to engage with the course material in new ways.

Throughout this chapter we have tried to illustrate that there is no single correct approach to the teaching of anything, including genetics. We hope that the complexity of contextual issues raised above will serve as a sobering reminder that selecting an instructional sequence should not be based on a course, textbook, or subject matter tradition alone. Instead, all instructional practice must be based on sound pedagogical principles that take into account both the structure of the discipline and the current intellectual state of the learner.

17

Collaborative Groups of Genetics Instructors

I am in the middle of midterm exam marking at the moment... what a disappointment. I honestly ache to see them SO lacking in their understanding of even the basic principles (e.g., that a mutant phenotype is the result of a NON functional gene not a functional gene!!). I mean... where can you go if they don't understand that? And how much time does it take to teach them before it becomes "not worth the effort?" ... Neither they nor we can afford to take four months saying the same thing over and over hoping that one day they will take the time to sit and think it through for themselves. I don't want to feel this way, but... After 300 exams (with 200 to go) I can hardly bear to pick up my red pen again.
Despondent e-mail from a graduate student teaching assistant, 1995.

It is likely that many of the people reading this passage have had experiences similar to those of the graduate teaching assistant above. Such feelings of despair, frustration, and self-doubt arise not only when marking, but in our lecture halls, tutorials, and during office hours. In these times of despair, ruminations boil down to questions such as, "What is going wrong?" "Is it me?" "Is it them?" However, these thoughts are part of the bigger and more fundamental questions such as "How should I teach?" "How much time should I spend?" and "What can I do if my students don't understand?" Like many struggles in life, in the end these problems have to be dealt with by individuals, but there is no reason why the search for solutions cannot be undertaken by collaboration with others of like mind. Indeed it can be argued that discussing teaching and learning problems in a group can provide not only consolation in times of doubt, but also a fertile environment for making proactive decisions leading to productive action. In this chapter we share our efforts at the University of British Columbia to move away from the solitary notion of teaching (in introductory genetics), and discuss what can happen when a group of instructors meet to discuss teaching on a regular basis.

Our story begins with our joint decision to set up a study group on genetics teaching. While our decision was based on a number of factors, the main factor was our realization that our discussions with each other on teaching and learning issues had been enormously enlightening and educational. Both of us had experienced learning to teach our respective subjects in isolation, and had discovered through our informal discussions that having a shoulder to lean on and a sympathetic ear were useful devices. Our conversations had led to a shared understanding of pedagogical problems, engendered creative instructional ideas, and bolstered a willingness to explore solutions and soldier on when things looked bleak. We reasoned that if having regular discussions worked for the two of us, then it might also be a powerful framework to explore with a larger group of university instructors. Because the teaching system in our large course depends heavily on graduate student teaching assistants, we decided that our Collaborative Study Group (CSG) should consist of professors plus professors-in-training (the graduate teaching assistants). Supported by a small grant from our University Teaching and Learning Enhancement Fund and monies from the Dean of Science we forged ahead and posted the following advertisement on the bulletin board in the genetics office.

Collaborative Study Group for Improving

University Genetics Teaching and Learning

Who Should Participate

You should consider participating IF:

- You respect and take your teaching seriously.
- You are interested in thinking about university/college teaching as part of your career options.
- You are a "professor-in-training."
- You are willing to critically examine instructional practice and explore options to improve your students' learning.
- You are comfortable discussing candidly your own teaching practices and your work with students.

- You are willing to open your classroom—and consider it a laboratory.
- You are interested and willing to contribute four hours/week (approximately) to the group for a $1000 honorarium. (Two-hour meeting/weekly + some other tasks or reading or observational time.)

Who Should Not Participate

You should NOT consider participating IF:

- You are seeking some additional money with minimal responsibilities.

To be honest, we were both worried that nobody would sign up. After all, we believed that most graduate students weren't planning to teach for a living—they were seeking academic positions where skills in research and publishing were the valued criteria of success. But our fears were unfounded, for six enthusiastic volunteers joined us, a university collaborative study group on genetics teaching was established, and our sessions began before the semester got under way.

At our initial collaborative study group (CSG) meeting the two of us explained our current ideas on the difficulties students experience with the learning of genetics, and we shared our views on the nature of learning and understanding (presented in other chapters). We discussed how the goal for this group was to think about "How should we teach so students will learn?" Thus although our focus was to be teaching, our objective was student understanding. We also established that an important goal for the group was to generate and document ideas that others could use to enhance their teaching of genetics because we felt it was important that our meetings would result in some tangible products, not just talk. After this initial meeting was over we fretted (in private) about whether we would get our professors-in-training to become equal partners in our joint venture. As it turned out, it didn't take long.

To provide the study group with some common teaching and learning experiences to discuss we proposed that it would be important to visit each other's classrooms. To break the ice Tony volunteered his big class sessions for our first team visit. And because it is easy to point fingers and pick holes in any lesson, we established that the purpose of observing teaching was to think about the instructional style and content from the students' point of view and to make suggestions about how student learning might be enhanced. This still made it easy for the group to be critical of the instructor (as we found out in our first attempt to discuss this lecture). But, as we had discussed the dangers of criticism explicitly and openly, it seemed to raise everyone's sensitivity to the issue of how feedback and comments might be perceived by the recipient. After our second observation and discussion, something seemed to "click," and "learning" rather than criticism became the major theme in our discussions from then on.

A second device that seemed to be important in stimulating discussion was our decision to ask each participant to send their thoughts regarding the important points from each meeting via e-mail to one of us so that a summary could be provided the following week. The thoughts and ideas on issues of teaching started to trickle in—filtered and distilled by each member of the group. Different individuals remembered and regarded different parts of the discussion as significant—but it was the writing process that seemed to help crystallize the ideas that emerged from our discussion of teaching episodes. Thus each week we had a collective summary of what had been learned. This gave our group a tangible "sense of progress."

Based on the collective wisdom and written contributions of the CSG participants, ideas started to emerge. The ideas were new to some of us, but more important they provided a common ground that we all agreed on. The ideas ranged from general pedagogical principles to debates about how to handle the teaching of specific genetics concepts. Examples of the ideas that emerged through our discussion are provided below (some of these have been elaborated on in other parts of this book).

1. **It is important to focus student thinking during lectures.**

The first idea stemmed from the observation that students spend a lot of time copying. Therefore it is appropriate to have a visual display of questions or problems on the overhead projector as students come into lecture. (In smaller classes this could be a handout.) The logic was that students could take these down right at the start and then during the lecture the instructor could use these as a source of discussion or tasks—at which point students can focus on the thinking associated with the activity rather than copying.

2. **Students' initial ideas are important for teaching.**

It is important that we investigate and become aware of students' entering ideas and use these to plan teaching in the tutorials. This can be done through tutorial quiz questions or through "written probes" of student ideas. Tests are also a useful source of student conceptions—but due to the infrequency of tests, learning about student ideas through tests usually occurs too late for remediation. (It is interesting that this idea arrived at by the group is one of the pillars of the constructivist movement in education.)

3. **Aiding students in understanding key concepts in lecture diagrams will improve student learning in lectures.**

When using overhead transparencies in lectures it is important to help students focus on the "signal" (the key points), otherwise they may attend to details that are less important, which constitute "noise" and may distract student attention. This approach suggests that learning is enhanced if fewer overheads are used, with more attention and time being devoted to each overhead.

4. **Diminishing ambiguity and focusing on "closure" will improve student learning in the genetics classroom.**

A lengthy discussion arose about ambiguity, and it was stated that people understand language differently. This is a significant problem in education in general and in genetics in particular, because students have to deal with the English language and different sets of symbolic and graphical representations. Further, it was generally recognized that a characteristic of genetics students is that they express their ideas and answers in a loose, confused, and ambiguous manner. It was felt that an important task of the teacher is to foster rigorous thinking and expression, and that going from ambiguity to clarity and making the implicit assumptions of a problem explicit was an important part of problem solving.

An alternative perspective on ambiguity was raised—namely that ambiguity has a role in teaching. If used thoughtfully (and in a carefully crafted situation) some ambiguity and speculation could be used to generate "constructive dissonance." An important issue that followed from this was that if speculation, brainstorming, or ambiguity was introduced to promote thinking, "closure" must eventually take place so that students don't leave in a confused state. It was acknowledged that these strategies require a trusting relationship between the instructor and students.

5. PloTS, a device for sharing teaching and learning ideas

A particularly powerful idea emerged during our third weekly meeting after consideration of how to design teaching activities in a manner that would allow them to be easily communicated to other instructors. It became clear that activities for students that would promote understanding of genetics problems were being designed and used by our CSG participants. However, when they brought these activities to share at our meetings, the content was clear but the advice on how to replicate the activity in another classroom got lost.

From our regular discussions on how best to share a teaching idea with others in our group a new instructional device was born, "The PLoT." PLoTs are Processing, Learning, and "other" Tasks that include a clearly defined genetics problem or task *plus* some pedagogical advice for the instructor on how best to undertake this task with a class of students.

Generation and discussion of PLoTs soon became one of the foundational themes of our weekly meetings. Study group members identified topics and concepts for which they felt PLoTs were needed. They created specific PloTs to fit these needs. These were then piloted in their tutorials, refined, and shared at meetings. Over time the group decided that PLoTs needed to have a fairly defined structure. The set of common elements negotiated and agreed upon include the following.

a) a catchy title

b) a list of concepts dealt with in the activity

c) a rationale

d) a concise statement of the pedagogical structure

e) a detailed description of the task

f) recommendations and hints for success

We are still working to refine our group's collection of PLoTs to share with others interested in promoting genetics teaching and learning. To provide a sense of what a PLoT looks like, at the end of this chapter we include one prepared by Michael Kyba, one of the graduate student members of our collaborative study group in 1997.

In closing, our experiences working in larger collaborative groups of genetics instructors have been exciting and productive. Not only has working together promoted a shift from teaching individually to teaching by team effort, but also in addition, it has demonstrated the power of collaborative interaction in design of creative teaching devices. We suspect that any similar effort can be rewarding, even if less formalized, such as group discussions over a glass of beer on Friday afternoons. Those who teach university science and particularly genetics have common concerns and problems, and we believe that the reiteration and consciousness-level raising of group discussions are intrinsically rewarding, personally empowering, and educationally productive.

Sample PLoT, reproduced by permission of Michael Kyba.

PLoT Title: Deconstructing and Reconstructing Pedigrees

Concepts: Pedigrees, Probability

Rationale: *Students have difficulty understanding probabilities of inheritance of alleles in pedigrees. They often develop the ability to solve pedigree problems without truly understanding the principles at work. This exercise forces them to address the principles of inheritance and statistics at work in pedigrees.*

Pedagogical Structure: Solve a pedigree problem. Then break the pedigree apart into minimal elements, assigning a probability of an allele to each element. Build the pedigree back up by first joining the elements and second, multiplying their probabilities together. Groups then build designer pedigrees, exchange these with other groups, and then try to solve them.

Task:

A. Start with any pedigree problem, solving it in groups or at the board or however you want.

B. Now get the class to "break the pedigree apart." Develop a conversation by asking some probing questions, and analyzing the students' responses.

<u>Sample Questions:</u>

1. What is the functional unit of the pedigree from the point of view of inheritance?

- A single mating with offspring.

2. Why is this the functional unit of inheritance?

- Alleles are inherited in steps through generations. One step takes an allele from one generation to the next.

C. To determine the types of functional units that the allele passes through in the solved question from above, have the students draw them independently. They should draw one diagram representing each step in the transmission of the allele in question, with each diagram having only two parents and the one offspring in question.

137

- Ask students to give allele designations to the parents, but for the offspring, instead of giving the actual allele designation, give all possible allele designations (considering phenotype) with the probabilities for each. Note that the probabilities must sum to one.

D. Ask students to determine how many theoretically possible building blocks there are. Have the groups draw them and assign probabilities to the phenotype of unaffected offspring, as shown in the diagram below.

p(carrier)	0	1/2	2/3	1	1
p(h. normal)	1	1/2	1/3	0	0

E. Now have the class "build" a simple pedigree, for the case where the likelihood of having an affected child is ¼ (examples are shown in the diagram below). Note that in real pedigrees, there is no information representing genotype (ask them why they think this is), and therefore any essential genotypes must be determined by the pedigree itself.

F. Ask if there are other pedigrees that satisfy the given probability (¼). Have the groups try to come up with an alternative solution, such as the one shown below.

G. Now modify the pedigree such that p(affected first child) = 1/8. Note that the probability given can be broken into factors, which can be represented by the "functional units" from above: $1/8 = (\frac{1}{2})(\frac{1}{2})(\frac{1}{2})$

H. Give each group a more difficult pedigree to generate (1/64, 1/24, 1/6, 1/18, etc.)

I. When your students are done, have them turn in a good copy, mix them up, and hand them back to different groups to "solve." Note that they will solve them much more quickly than they can generate them!

This exercise takes most of the period. It is therefore worthwhile to incorporate the pedigree from an assigned problem into it.

18

Ethics and Eugenics

Genetics seems to have more than its fair share of ethical controversies. Some of the current ones are the production and use of transgenic plants and animals, cloning mammals (especially humans), documenting human genomic diversity, DNA typing and privacy, and of course eugenics. Do these issues belong in the classroom? We think so.

The medieval English poet John Donne put it well: "No man is an island. Send not to ask for whom the bell tolls, it tolls for thee" (Donne, 1624, p. 538). No human activity exists in a vacuum, devoid of social context or consequence. Science, which dominates all our lives, requires its own context. To strip it of this context leads to popular misunderstanding and fear of science and the misuse of science, both of which are detrimental to society. Yet few science courses in post-secondary settings (genetics courses included) pay much heed to ethical matters. Perhaps there will be a fleeting mention of a few issues, but rarely an in-depth analysis of the philosophical issues. Sometimes universities offer a special course in scientific ethics. Unfortunately this action may actually be counterproductive in that it enhances the view that the practice of science and the ethics of science are two different matters.

Genetics textbooks also give ethical issues little attention. Part of the problem is that textbook reviewers tend to be highly suspicious of ethical discussions, in particular those that may question or impinge on their own beliefs. Curiously, genetics texts that cater to non-science majors often do have serious discussions of ethics. Thus, the point is made clearly—ethics is not the domain of scientists and does not belong in courses that are aimed at those individuals planning to take on science as their profession. We disagree with this point of view. Clearly, scientists *are* part of the bigger society and the work of scientists ultimately impacts all of society. The "educated scientist" needs to be aware of and sensitive to the economic, societal, medical, political, moral, and ethical implications of his or her work.

Many issues that impact society can be linked with the curriculum of our genetics education courses. In this chapter we illustrate this point by focusing on one current ethical controversy. Our goal is to demonstrate the educational potential of uniting social, ethical, and scientific issues into the arena of genetics teaching. For our discussion we have selected the ethical controversy surrounding the Maternal and Infant Health Care Law (MIHCL) of the People's Republic of China passed in 1994 and perceived by some to be eugenic in nature. Eugenics in general has had a turbulent history ever since its conception in the late nineteenth century. The social acceptability of eugenics has waxed and waned over the decades. Likewise the scientific basis for eugenics has been hotly debated. Passage of the MIHCL provides a modern context for discussion of both the social and scientific aspects of eugenics.

Prompted by pressure to do something about the large number of people in China born with genetic handicaps, the Maternal and Infant Health Care Law was formulated "to improve the quality of the newborn population" (MIHCL Article 1). The MIHCL was originally labeled a "eugenics law" by those who prepared this statute. In their framing of the law and its principles the writers used the Chinese word "youseng" which means "healthy birth," and had looked up the word in a Chinese-English dictionary where the translation came out as "eugenic." As it happened, this translation was reasonably accurate because the original etymology of the word eugenics means "good birth."

The international reception to this law was entirely negative, prompted by the terrible track record of eugenics over the past century. But the question arose: Was the law eugenic in any negative sense? The answer to this question is a complex one as it depends on the interpretation of the science and the policies written into the law. The law contained many non-controversial articles that concerned the health of both mother and child. It also included a number of articles regarded to be contentious. The four most

141

controversial articles are cited below (translation provided by the government of the People's Republic of China).

Article 10

Physicians shall, after performing the pre-marital physical check-up, explain and give medical advice to both the male and the female who have been diagnosed with certain genetic disease of a serious nature which is considered to be inappropriate for child-bearing from a medical point of view. The two may be married only if both sides agree to take long-term contraceptive measures or to take ligation operation for sterility. However, the marriage that is forbidden as stipulated by the provisions of the Marriage Law of the People's Republic of China is not included herein.

Article 16

If a physician detects or suspects that a married couple in their childbearing age may suffer from genetic disease of a serious nature, the physician shall give medical advice to the couple, and the couple in their childbearing age shall take measures in accordance with the physician's medical advice.

Article 18

The physician shall explain to the married couple and give them medical advice for a termination of pregnancy if one of the following cases is detected in the pre-natal diagnosis:

    1) The fetus is suffering from a genetic disease of a serious nature;

    2) The fetus is with defect of a serious nature; and

    3) Continued pregnancy may threaten the life and safety of the pregnant woman or seriously impair her health due to the serious disease she suffers from.

Article 19

Any termination of pregnancy or application of ligation operation shall be agreed and signed by the person concerned. If the person has no capacity for civil conduct, it shall be agreed and signed by the guardian of the person.

There has been considerable disagreement by ethicists, scientists, physicians, and politicians over whether or not the phrasing of these articles reflects eugenic intent or not. Some saw the wording as innocuous, some saw it as similar to accepted practice in other countries, and some saw in it an intent that was a violation of the human right to reproduction. Here, the context of the reader seems to be paramount. Part of the issue has been the definition of the word eugenics. It is evident that this term means different things to different people. The term was originally coined by Galton (1883) who defined eugenics as "The study of agencies under social control that may improve or impair the hereditary qualities of future generations of man, either physically or mentally."

This definition is open to several possible interpretations, which is not surprising because it was drawn up before genetics became an established discipline, and before the laws of population genetics were deduced. Perhaps the main issue is whether the intent of eugenics is to systematically change the genetic structure of populations (i.e., to change disease allele frequencies) or to merely change the frequency of genetic disease in the population.

At the Eighteenth International Congress of Genetics held in Beijing in August 1998, a special international workshop was held on the topic of the theory and practice of eugenics in general, and specifically regarding the MIHCL. The debate that took place during this workshop illustrates the educational point raised earlier in this chapter; namely that context is a crucial component of science. The conclusions that emerged from the debate illustrate the breadth and the complexity of the issues, as well as the value of scientists looking beyond the lab and considering the ethical issues related to their scientific findings. There were eight conclusions, which go considerably beyond the topic of eugenics and are applicable to a wide range of ethical dilemmas. These conclusions can be used as a framework for initiating classroom discussion of the ethical implications of genetics research.

1. Countries share many ethical principles based on the will to do good and not harm. These principles can be applied in many different ways.
2. New genetic technology should be used to provide individuals with reliable information on which to base personal reproductive choices, not as a tool of public policy or coercion.
3. Informed choice should be the basis for all genetic counseling and advice on reproductive decisions.
4. Genetic counseling should be for the benefit of the couple and their family: It has minimal effect on the incidence of deleterious alleles in the population.
5. The term "eugenics" is used in so many different ways as to make it no longer suitable for use in scientific literature.
6. In formulating policy on genetic aspects of health, international and interdisciplinary communication should be carried out at all levels.
7. It is the responsibility of policy makers concerned with genetic aspects of human health to seek sound scientific advice.
8. It is the responsibility of geneticists to educate physicians, decision-makers, and the general public in genetics and its consequences for health.

Each one of these conclusions could serve as a starting point for classroom-based examination of ethics in science generally, and in genetics specifically. A few comments (numbered to correspond to the above list) expand on the terse prose of the conclusions.

1. This makes the point that one nation's ethical actions might be another nation's evil. We are urged to overcome the temptation to be judgmental based on our own codes of behavior, and attempt to understand the cultural context of differences.

144

2. Any new technology carries immense power, which can be abused. This item makes the point that the ultimate use of science must be to promote personal freedom.

3. The phrase "informed choice" was key here, and was selected over "informed consent" (which might be coercive). Informed choice can only come from understanding, which in turn is a result of education.

4. Here any notion of the application of eugenics to "improve" the genetic structure of a population is rejected. In its place, genetic counseling is seen as applying genetic insight for the benefit of the family.

5. As mentioned previously, the term eugenics has no universally accepted meaning. Any term that has no universally accepted meaning has no place in science.

6. This is a principle of internationality. In the global village we are all citizens of each other's countries, so lines of communication are needed for the purpose of consultation on any application of science/genetics to the human condition.

7. This is another point concerning lines of communication. In decisions about the application of scientific principles, scientists must be consulted.

8. Geneticists cannot blame any but themselves if society does not understand genetics. We are the ultimate teachers. It is our responsibility to put our subject across to the public and to all who are affected by the discoveries and application of our subject.

Some type of eugenics has been proposed or applied in most countries, so in addition to the MIHCL of China, there is ample opportunity for students to link social and scientific aspects of eugenics in their own community. Another challenge would be to consider whether eugenics is being carried out anywhere in the world today. (Is "ethnic cleansing" a type of eugenics?) Of course, the impact of the availability of DNA markers on the scientific feasibility of eugenics would also be a good topic for debate.

Bringing ethical issues into the science classroom may seem a frightening prospect, as there are no black and white answers to ethical questions. The fundamental point to keep in mind is the educational purpose of discussing ethical problems is to inform and enlighten students about the issues, not to indoctrinate or praise a particular perspective.

This point needs to be made clear to students before you begin to discuss any issue where ethical points of view may differ.

Many possible strategies exist that will get students thinking about the ethical implications associated with scientific knowledge. A simple structured teaching approach might begin with illustrating the multiple perspectives that need to be examined when considering the impact of scientific findings on society. These perspectives include scientific, economic, medical, political, moral, and ethical points of view. To provide concrete practice, have students identify these perspectives after reading a controversial news article in which many perspectives are being reported. A second type of strategy would be to have your students critically examine a specific research finding and identify any risks or consequences associated with applying these findings uncritically. A final activity would entail selecting a controversial issue such as the MIHCL and assigning students to any one of a number of legitimate alternative perspectives, asking them to argue the issue through the lens you have assigned them. This type of activity is ideally suited to a debate format.

One practical and familiar aspect of genetics you could begin with would be to have your students compare the approaches of eugenics with those of genetic counseling. How do the approaches and goals differ legally, medically, scientifically, and ethically?

Still wondering how to start? Why not begin the conversation with a debate about current practices in post-secondary science courses. Teaching about genetics without considering the ethics of genetics is a controversial issue in itself. Ask your students what they think.

Reference

Donne, J. (1955). Devotions upon emergent occasion. Meditation XVII in John Hayward (Ed.), *Complete poetry and selected poetry*. London: Nonsuch Library (original work published 1624).

19

The Geneticist as Dr. Frankenstein

or

Public Education in Genetics

Not only is the general public poorly informed about genetics, but also in the public's eye geneticists have been singled out among scientists to be cast in the role of Dr. Frankenstein. Like Mary Shelley's famous doctor, the geneticist is seen as dabbling in the unnatural, creating various types of monstrous products that are perceived as harmful. Now, in another scene lifted straight from the book, the villagers are rising up with their pitchforks and storming the castle. This view of genetics is having serious consequences on the practice of our discipline, partly because of the negative image that is so widespread, but also because of the current unacceptability of certain types of research, especially those involving transgenic modifications.

The evidence has been accumulating slowly, but geneticists seem to have woken up to the problem only recently now that the hordes are actually at the castle gates. During the early days of work on recombinant DNA there was much discussion over the safety of the techniques, resulting in public information meetings and debates, but then things seemed to settle down and geneticists relaxed. The late 1980s saw the publication of surveys done in Canada, the United States, and Europe, showing that the public was surprisingly ignorant of fundamental scientific issues. However, these reports did not set off many alarm bells for genetics instruction specifically. A few people crusaded against the introduction of products such as ice nucleation bacteria and Flavorsaver tomatoes, but again these were treated as storms in teacups by most researchers. For us (the authors), two incidents acted as a wake-up call that announced that things had degenerated badly.

First, in a widely publicized case, the so-called Unabomber sent a letter bomb to the office of a prominent geneticist as a protest against the new genetics. Second, photographs appeared in papers around the world showing members of the Greenpeace organization engaged in a large and well-orchestrated protest in Liverpool harbor involving flotillas of Zodiacs, huge banners and projection of slogans on the sides of ships. Their target was freighters delivering transgenic soybeans from North America to Europe. Their banners screamed "Stop genetic pollution." For Vancouverites, this was doubly stunning as Greenpeace was born in this city, where they were well known and respected for their protests against whaling and nuclear weapons. What a shock it was to learn that for Greenpeace now evil geneticists had replaced the evil whalers and nuclear powers.

Most people know that in the last few years the pace of protest has quickened and embraces most of the developed world. Now huge anti-GMO parades are a common sight, and we even have a new brand of terrorism called ecoterrorism, the aim of which is to oppose GMOs, often indulging in unlawful destruction of transgenic plants to do so. The front cover of a recent *Economist* summed it all up with a drawing of a "Frankenfood" potato exclaiming "Who's afraid of GM foods?" Despite the news that many GM crops seem to hold out obvious benefit for the world's poor and hungry (high-carotene and high-iron rice, resistance to pathogens, vaccines, longer shelf life, reduced allergens, for example), and despite the announcement by the esteemed scientists of the U.S. National Academy that there is no scientific evidence of harmful effects of GM foods, our breakfast reading continues to assail us with news of fresh protests, boycotts, and incidents of vandalism. Both health and environment are perceived to be under threat from "genetic pollution" that is "unpredictable, uncontrollable, unnecessary, and unwanted" (Greenpeace).

Human and mammalian geneticists have not escaped similar criticism. The cloning of Dolly the sheep and other agricultural animals has led to protests against the way in which geneticists are allowed to emulate Dr. Frankenstein or even "play at being God," and has raised the specter of human cloning, which seems to be universally loathed. Patenting of genetic material also seems generally to produce outrage. Human DNA seems to have found a special niche in the matter of individuality, and people who do not object to fingerprinting using actual fingers, find DNA fingerprinting to be an unacceptable invasion of the persona. For possibly the first time ever, insurance has become a controversial topic because of the possibility of DNA markers showing predisposition to disease and death, although other types of information on predisposition are widely used. People who are willing to give up a certain amount of personal freedom and individuality for the common weal, and who are interested in and approve of genealogy, seem loath to divulge the details of their genetic makeup.

How did this happen? Who is responsible? To quote Pogo, "We have seen the enemy and he is us." That is, we have nobody to blame for the present bad image of genetics other than ourselves. A fine mess we have gotten into, and it is caused largely by our inattention and failure to educate the public in genetics. Let's consider a few points that might throw light on how this failure has occurred. Here we strive not to take sides on the rights and wrongs of GMOs and other genetic "monstrosities," but attempt to unravel some of the educational components.

1. Public education in science stems largely from the education that takes place in science faculties.

It is true that public lectures and colloquia, visits by scientists to schools, TV programs, and popular literature all contribute to public education in science. A few good examples from the media are David Suzuki's TV show "The Nature of Things," and many items on the Discovery Channel, PBS, and BBC, all of which aim at not only highlighting the findings of science but also the nature of scientific curiosity and research. Children's shows, such as "Bill Nye, the Science Guy," tend to deal in "golly gee" science told to the children by an entertaining scientist and his helpers. Although engaging, these children's programs show little about the way that science is actually done. In the popular literature, great science (including genetics) articles regularly appear in *Scientific American, The New York Times, Time* magazine, and *National Geographic.*

However, all of these efforts put together pale in comparison with university science education. One reason is, of course, that most scientific advances are made in these institutions so they act as leaders in providing cutting-edge methods and results. But the main reason is that the trickle-down effect of university science education is immense simply because science faculties educate school science teachers, who not only educate students headed for university but also educate every member of society. Hence the potential for maximum leverage is at this point, so if there has been failure it has been at this stage. It follows that if changes are to be made, they too must be initiated at this stage.

2. Context is all-important in genetics learning.

There is a joke in the real estate trade that the three key elements for making sales are location, location, and location. There should be a similar saying in science education: the three keys to learning are context, context, and context. Without context, teaching and learning become a game played by both the teacher and student (see Chapter Eleven). With proper context, learning is spontaneous and the instructor merely needs to facilitate the students' efforts. Most of the failures of science education generally can be attributed to our failure to provide an adequate context for the educational experience. More specifically in the present connection, most of the controversy and suspicion surrounding genetics in the popular eye can be attributed to a lack of suitable context for genetic advances. Some examples follow.

a) Genetic technology is like any other technology. All technology is "unnatural." All technology has positive and negative attributes. Whether true or not, most people would agree that the Industrial Revolution of the nineteenth century, based on scientific advances in physics and chemistry, has been a success overall because it has increased the standards of living in those countries where it took place. However, the downsides are also immense. Take one of the products of the Industrial Revolution as an example: the motor car. Beloved by most for the freedom it provides, motor cars are yet a massive source of ecological and health problems including air pollution, mining for metals and oil, and deforestation and loss of biodiversity due to the planting of rubber trees for tires. Undoubtedly there will be some negative aspects of biotechnology, specifically GMOs. The potential bad of genetic technology must be weighed against the potential good. Similarly, genetic technology must be stacked up along other technologies, not singling out genetics for special treatment or bad publicity. The chemical revolution of the previous century ("better living through chemicals") has given us plastics, dyes, and many other useful compounds but it has also given us the hole in the ozone layer, chemical pollution of most natural bodies of water on the planet, radioactive contamination, and thousands of festering toxic lagoons around the world. These downsides are tolerated (admittedly partly from ignorance) but in comparison the potential hazards of genetic technology seem relatively minor.

b) In the genetic technology debate, industry is painted as villain. Yet industry is merely doing the same things in genetics that it does in any other endeavor, that is maximize its profit from some competitive edge that it can garner. So the

151

potential injustices of patenting genetic information are not different in principle from other types of patenting, and the suffering of farmers from terminator corn stocks are no different from the suffering from any other type of protection of marketable ideas. Furthermore, industry can indeed abuse human rights, but this is a property of industry in general, which our political and economic systems have bought into, not only of genetic industries.

c) Genetic engineering has been branded as unnatural. Indeed it is, but, as professed above, that is the nature of technology in general. Technology is an intervention into nature, into the natural state of things. Transgenic corn is no more unnatural than the multitude of mutant forms of plants used in gardens throughout the world; but whereas one delights, the other horrifies. Gene therapy is no more unnatural than therapy by drugs. Together, genetically engineered organisms are no more unnatural than plastic running shoes or television sets.

A better context for the "unnaturalness of GMOs" is the simple fact that many of the items we rely on in our lives have already been subjected to genetic improvement over the past 100 years. Sheep, cattle, pigs, and poultry have all been modified to adjust their nutritive or other properties to suit our preferences and perceived needs. Plants similarly have been bred for food and other properties. Indeed genetically modified plants now form the basis of the world's food supply (notably wheat and rice). The same geneticists who were trusted for the design of these products are now surprisingly under suspicion for the design of GMOs.

d) The potential ecological and health impact of artificial agricultural products is an important issue. Yet here again is a double standard because technologies other than those of molecular genetics are now wreaking havoc on the planet's habitats, but currently enjoy a relatively low profile in comparison to that of genetic manipulation. For example, the much quoted results on the potential adverse effects of Bt toxin-containing pollen on Monarch butterflies pale before the habitat destruction that threatens the migrations of this same species. Invasive species all over the planet are causing numerous ecological crises right now, all across the planet. Will transgenes be transferred into wild species? Perhaps, but because a large proportion of the world's land is currently planted with specially bred crop species not native to the region, there is already ample opportunity for this type of gene spread now. On the health front, the National Academy of Sciences' report notwithstanding, it is perhaps the ultimate irony of context that Canada's and America's biggest potato growers, who are the suppliers to the multinational fast-food chains, have decided to withdraw all transgenic stocks on the grounds of possible negative effects on health, while apparently remaining unmoved by the documented evidence of real and massive health risks from the fat and cholesterol in the products of these chains.

e) The point is often made that geneticists don't know what they are doing when they add a transgene to a recipient genome. A single "stone" is thought to possibly cause a mighty "ripple." This is mysticism. Many such stones have been dropped into numerous genetic ponds. Conventional breeding depends on mutants, most of which are pleiotropic in their effect, yet no man-killing monsters have emerged yet. Furthermore, what could be more unnatural than hybrid plants, genomic blends that could not have happened in nature? Take for example the amphidiploid Triticale, a juxtaposition of the genomes of two different genera, wheat and rye. It seems that enough experience has been gained from one hundred years of genetic research to be pretty confident that changing a gene or two is not

153

going to produce a new kind of genetic time bomb. Thus the accumulated experience of research provides the context. For another context, we can again make a comparison to parallel technologies: that is, do we know the long-term health effects of most newly developed drug treatments, or even vitamin supplements? And what about the long-term effects of TV and graphic videos on children or cellular phones on adolescents and adults? The answer is that we do not know this impact even though it seems inevitable that it is immense and detrimental, yet these technologies are widely embraced.

It is not our point to offer *apologia* for genetic modification. One cannot logically deny the possibility that there will be some negative aspects. The point is that we must keep the issue in perspective by comparison with other technologies. The same is true for testing. Most agree that testing the impact of new technologies is a good idea. There have been too many major catastrophes for this not to be self-evident (DDT, fluorocarbons, and PCBs come to mind). However, *all* technologies need to be tested, including those that are currently far more menacing than genetics. Apparently new potent chemicals are being developed and used faster than they can be tested.

What about solutions? Ideas that are deeply entrenched in a culture are the most difficult to change, and it looks like the ideas about genetic manipulation are now indeed firmly embedded. One answer that seems appealing to industry is simply to wait for the storm to pass. However, from the point of view of practicing geneticists, what are our options? We, the authors, don't pretend to have the answer to such a profound question that seems to require a revolution in educational methods, but here are some things we do believe.

1. If anything is to change, the change must originate in the university and college settings.

2. Because the issue is not merely that of GMOs and cloning, but of technology in general, the solution cannot come from genetics alone, but must come from concerted and systemic change throughout the whole educational system.

3. If any change is to be accomplished, it must hinge on educating students squarely within an appropriate context for science. The connections between scientific knowledge, scientific research, cultural values and needs, and even the arts, become paramount in any restructuring that might occur. Even though science is often *done* in an ivory tower, science must be taught as the integral part of our cultural fabric that it really is. In this way the consequences of scientific knowledge should become a central part of the teaching system.

4. Such re-contextualization must start early in the education system, at the lower school grades.

5. Because biology has the advantage over most of the physical sciences in that the questions of biology stem from observation of the natural world of organisms and habitats, which nearly all young children find interesting and appealing, this seems to be an appropriate and powerful source of context. Biology can begin just outside a student's back door, in the garden, in the community, which teems with life, even in the most urban environments. This is where the revolution might well begin, by introducing young people to this fascinating world that surrounds them, and promoting a life-long love of it.

THE VICIOUS CYCLE

UNIVERSITIES COLLEGES

SCHOOLS

educate science teachers

educate science students

20

Wrap-up

Scattered throughout the book we have offered several activities to improve the teaching and learning of genetics. In this final chapter we provide for convenience an "Executive Summary" of strategies introduced earlier plus some additional ones that you may wish to try with your classes.

1. **Quick Starts**

   To get the students warmed up and in the mood for active participation, a simple problem is put on the overhead for them to attempt to solve while people are coming in and taking their seats. The problem may be directly related to concepts and issues being dealt with later in the class or it may serve as a quick review of the previous week. How you proceed after they complete the problem is flexible (except of course we would suggest you don't simply "show them how" to do the problem.) Do keep in mind this is a "starter" activity and should not take up the entire class period.

2. **Models and Mimics**

   The traditional method of modeling (i.e., demonstrating) the way to solve problems is useful as an illustration but cannot by itself teach problem solving. However, this approach can be modified to allow student participation. The instructor *models* how to reach the solution to a problem on the board, and a student is selected at random to pass the chalk and attempt to *mimic* (i.e., repeat) the solution modeled by the instructor. At any stage the student can hand on the chalk to another student of their choice, who then takes over the solution. Because the solution is given, this method overcomes (to a large extent) the fear of speaking in public, but forces the student to actually go through the mental steps of the solution.

3. Dyads

This is an instructional grouping strategy that has been around for some time, but we have found it useful for generally encouraging students to discuss data. It gets around the "silent class syndrome" (i.e., no one will answer the question posed). Students form pairs (dyads), are given the problem or partial problem, and after two to three minutes people are asked to speak on various aspects of the solution. Having done some processing (and practicing) in the discussion with their dyad partner, the student is more inclined to speak and discuss publicly. By first working out one's ideas with a partner there is less risk of being individually regarded as "wrong."

4. Think Aloud Pair Problem Solving (TAPS)

This method, which is an extension of the dyad approach, gives the partners defined roles. One of the pair is the "problem solver" and the other is the "listener." Each has an active role. The solver attempts to solve the problem out loud (while writing or illustrating with diagrams or doodling if necessary), articulating all the logic, thinking, and principles behind the steps taken. If the solver cannot continue he or she describes why they are stuck and what is needed for them to be able to proceed with the solution. The listener "keeps the solver honest," asking such questions as "Why did you do that?" "What principle did you invoke there?" and "Why didn't you multiply by this instead of that?" The listener attempts not to solve the problem or give big hints. After a suitable interval the class discussion is held on the processes discussed, focussing as much as possible on the steps and not the right answer.

5. Think Aloud Problem Solving in Triads

This is a modification of the TAPS method with the addition of a third person, the scribe, whose job it is to record in writing the path taken either to the solution or to the dead end. The addition of the scribe makes it easier to report to the rest of the class the results of the discussion within the TRIAD.

6. **Unraveling Genetics Nomenclature**

Take symbols for wild and mutant allele symbols from different organisms and try to interconvert. Draw a diagram of what a symbol means. Draw diagrams that show the meaning of a slash (/), period (.), comma (,), colon (:), double colon (::), etc.

7. **Moving Between Multiple Representations**

Genetics is riddled with multiple representations. For example, genetic crosses can be represented by pedigrees or Punnett squares, Xs represent metaphase chromosomes, X-chromosomes, crossovers, the symbol of a cross, etc.; a chromosome can be represented by a line, parallel lines, a "sausage" structure or an X; a DNA molecule can be represented by a line, a double line, a ladder-like structure, chemical symbols, or a double helix. Rather than simply mentioning to students that this can cause confusion, exercises can be designed to help students keep the different representations clear in their thinking. For example, students can be challenged to move from one representation to another, as in converting information from a Punnett square into a pedigree. Multiple representation exercises can be presented regularly throughout the term.

8. **Zooming**

Experts in a given field are able to "zoom" in and out easily between different levels of organization of a subject. For example, in the case of genetics we move quickly from discussing DNA to chromosome to cell to organisms, etc. This requires understanding the connections between the various levels and many students have great difficulty with this. Exercises can be designed in which students are challenged to zoom from one level to another. For example, starting with a human genetic disease phenotype, the student can be asked to describe what is going on at the organ level, the cell level, the protein level, and the DNA level. Sometimes it is helpful to ask the students to think about what is going on from a different perspective or to use a different means of explaining the various connections between the levels. Translation activities can be used to do this (see next item).

9. Translation Activities

Translation activities can take many forms, but in all of them the student has to translate (transform) conceptual information presented in one form into another. For example the students may "translate" notes into cartoons; diagrams into a story; graphs into poetry (e.g., tell the story of an oxygen molecule that is inhaled by an athlete). Translation tasks often require the learner to communicate or explain concepts to some kind of "audience" described in the activity. In order to do the "translation," students must take the concepts, principles, rules, or definitions and think deeply about what they mean. Through the process of "translating" the learner will need to think about and examine what they know and (hopefully) identify what they do not understand.

10. Concept Maps

Invented by Novak and Gowin (1984), the concept map has great application in genetics because of the difficulty students have in relating one area of the subject to another. For example they might "understand" mutation, and they might "understand" modified Mendelian ratios, but in an exam they cannot fit the two concepts together. There is also the problem referred to under zooming (see above), which is the inability to interrelate the different functional levels of the genetic process. In a concept map the student is given a half a dozen or so genetic terms and asked to write them down and draw lines connecting the related terms, writing on each line drawn a statement that explains the nature of the connection. The exercise seems simple and trivial but it is remarkably powerful at revealing areas of weakness and uncertainty. Even professionals who have tried to draw these maps are often amazed at their inability to properly describe the relationships. It must be stressed that there is no "right answer"—the map is just an aid to thinking and constructing a better understanding of the connections between the concepts.

11. Practical Challenges

Take any genetic entity such as an aneuploid, arginine mutant cosmid library, etc., and devise a set of detailed step-by-step experimental procedures whereby one could make one of these. Include details of where to obtain materials.

12. The "Formula" Approach to Problem Solving

This approach is to try to reconstruct some of the steps that skilled problem solvers go through in attempting a problem. It consists of a series of linear and sequential steps that narrow down the domain of genetics that the problem is in. It is a type of taxonomic key for the components of genetics. Some of the steps may seem obvious, but placing the problem in context is one of the more difficult things for students, and they often end up trying to force-fit a solution that reveals they have not placed the problem correctly into the web of genetics. Different people will have different series of steps; students should develop their own. Here is one linear sequence:

What organism are we dealing with? → Are we working with a stage that is haploid or diploid? → Mitosis or meiosis? → One gene or more? → How many alleles? → Autosomal or sex linked? → Dominant or recessive? → Linkage or independent assortment? → Gene interaction? → Any evidence for mutation? → Any evidence for a change of chromosome number or arrangement? → and so on.

13. Problem Expansion

A genetics problem is only the tip of an iceberg of understanding. The words and sentences used in setting the problem subtend huge areas of genetics that a student must know in order to be able to attempt a solution. The problem expansion activities are designed to E X P A N D or "blow up" a problem by asking students multiple peripheral questions about the problem focusing on the problem itself rather than the solution to the problem. This helps clarify the underlying assumptions made by the problem setter and gets the students into the habit of looking beneath the top layer of the problem.

14. Problem Setting

Getting students to participate in problem setting (making up problems) provides students with another and different view of the problem-solving challenge, and provides insight into the kinds of items that are required (and must be specified) in order to analyze genetic data. There should be some structure for this, so as not to leave students all at sea. Two structural approaches are:

- Present some experimental data from a paper or a book and ask the students to make up a problem out of it. Ask other students to solve the problem, and to point out any weaknesses in the problem.

- Devise a problem that is the reverse of another problem. Generally genetics problems are of two types: (i) Forward working problems: give genotypes of parental organisms and ask for predictions of outcomes of crossing them, mutating them, etc. (ii) Backward working problems: give experimental results and from this deduce the parents or other starting material. Students are given one of these types of problems (you can provide either a solved or an unsolved problem) and then asked to convert it into a problem working in the reverse direction.

15. Writing about the Writing

As stated above, part of the problem in learning genetics or any other subject is the tendency to allow material to float through the brain without processing it. Writing on the writing or producing "Commentaries" challenges the student to think about, analyze, and comment upon a specimen (e.g., a "genetic" corn cob), a photograph (either from the book or presented in class), a color slide, or any other display (such as a layout in the text). The commentary can be spoken, or in some cases written (writing on the writing).

16. Debates

For some controversial topics such as GMOs, DNA fingerprinting, eugenics, fetal testing, patenting of gene sequences and strains, etc., it is useful and fun to pick teams for a semi-formal debate. The two teams are given five to ten minutes in class to prepare, and then proposers and opposers to the motion alternate. The instructor tries to wrap it up with a debriefing.

17. Scrutinizing Solved Problems

Worked examples attempt to show the reader how to solve a problem and most texts contain a collection of these "solved problems." Theoretically this should be helpful because such examples clearly lay out the steps that need to be taken to get to a solution. Often these steps are the bare bones of the solution. The author has not communicated directly the thinking behind each step. And, most often students don't really analyze the thinking and procedures that are contained in each step of the solution. Many students think that reading through these worked examples will help them, and for some it does. But it isn't the reading alone that helps, it's the mental interaction that takes place between the reader and worked example that leads to understanding. This mental "scrutinizing of worked examples" is another metacognitive behavior—one that occurs subconsciously and spontaneously for some learners and not for others. One way to develop skills in scrutinizing worked examples is to be *explicit* about the need to do so and to provide *tasks* which get students to engage in this activity publicly. This strategy can be combined with other structures such as DYAD or strategies such as Models and Mimics.

18. Playing the Role of Genetic Counselor

In most texts the instructor can find human "genetic dilemmas" of the type that confront genetic counselors. Because many students are interested in medicine, these situations are excellent "hooks" on which to hang learning. Students are given the dilemma on a handout or overhead, and asked for advice. After a short period of writing and thinking, the instructor asks for ideas, and later debriefs the session. The approach comes close to the formal type of "problem-based learning" that is used in many medical schools. Many of the dilemmas are very "genetics problem-like" and the connection is easy to make.

19. Talking Genetics—Using Fun and Games

Occasionally more creative strategies may be presented to get students talking genetics. Activities such as the examples below give students practice with the language of genetics and help to clarify for them (and for the instructor) when concepts are clearly understood or fuzzy.

- Ask two students to ad lib an imaginary discussion between Gregor Mendel (the "father of genetics") who did not know about chromosomes, and a microscope maker who had just witnessed the behavior of chromosomes at meiosis.

- Play the old radio game "Got a minute?" with genetics terms. Students have to speak for one minute without hesitation or pause on such topics as "frameshift mutations" or "intron splicing."

We hope the general idea comes through from this list of activities. Many of them require risk-taking not only by the student, but by the instructor too. Many appear to be difficult to apply in large classrooms. However, whereas it is tempting to use this as an excuse for not trying them, these strategies have been used in large classes quite successfully.

Does all this work? Do students emerge better-educated in genetics? We have found that the students who adopt the constructivist agenda claim that it has changed their approach to science in general. A large study has been done in physics to test the efficacy of student-centered education and the results showed spectacular success. We are confident it is the way of the next millenium. However, as with any conceptual change, and the institutional change that must accompany it, the going can be rough not only for the students, but the instructors and teaching assistants too. Many instructors find it difficult to give up the role of expert. Many students resent being asked to do tasks that do not seem to them to be "real education." Whatever the results, we guarantee an interesting experience that should produce a new view of genetics education.